Rufus Phillips Williams

Introduction to Chemical Science

Rufus Phillips Williams

Introduction to Chemical Science

ISBN/EAN: 9783337034429

Printed in Europe, USA, Canada, Australia, Japan

Cover: Foto ©berggeist007 / pixelio.de

More available books at **www.hansebooks.com**

INTRODUCTION

TO

CHEMICAL SCIENCE.

BY

R. P. WILLIAMS, A.M.,

INSTRUCTOR IN CHEMISTRY, ENGLISH HIGH SCHOOL, BOSTON,
AND AUTHOR OF "LABORATORY MANUAL."

BOSTON, U.S.A.:
GINN & COMPANY, PUBLISHERS.
1894.

Entered according to Act of Congress, in the year 1887, by
R. P. WILLIAMS,
in the Office of the Librarian of Congress, at Washington.

TYPOGRAPHY BY J. S. CUSHING & CO., BOSTON.
PRESSWORK BY GINN & CO., BOSTON.

PREFACE.

THE object held constantly in view in writing this book has been to prepare a suitable text-book in Chemistry for the average High School, — one that shall be simple, practical, experimental, and inductive, rather than a cyclopædia of chemical information.

For the accomplishment of this purpose the author has endeavored to omit superfluous matter, and give only the most useful and interesting experiments, facts and theories.

In calling attention, by questions and otherwise, to the more important phenomena to be observed and facts to be learned, the best features of the inductive system have been utilized. Especially is the writing of equations, which constitute the *multum in parvo* of chemical knowledge, insisted upon. As soon as the pupil has become imbued with the spirit and meaning of chemical equations, he need have little fear of failing to understand the rest. To this end Chapters IX., XI., and XVI. should be studied with great care. In the early stages of the work the equations may with advantage be memorized, but this can soon be discontinued. Whenever symbols are employed, pupils should be required to give the corresponding chemical names, or, better, both names and symbols.

The classification of chemical substances into acids, bases and salts, and the distinctions and analogies between each of these classes, have been brought into especial prominence. The general relationship between the three classes, and the general principles prevailing in the preparation of each, must be fully understood before aught but the merest smattering of chemical science can be known.

Chapters XV.-XXI. should be mastered as a key to the subsequent parts of the book.

The mathematical and theoretical parts of Chemistry it has been thought best to intersperse throughout the book, placing each where it seemed to be especially needed; in this way, it is hoped that the tedium which pupils find in studying consecutively many chapters of theories will be avoided, and that the arrangement will give an occasional change from the discussion of facts and experiments to that of principles. In these chapters additional questions should be given, and the pupil should be particularly encouraged to make new problems of his own, and to solve them.

It is needless to say that this treatise is primarily designed to be used in connection with a laboratory. Like all other text-books on the subject, it can be studied without such an accessory; but the author attaches very little value to the study of Chemistry without experimental work. The required apparatus and chemicals involve but little expense, and the directions for experimentation are the result of several years' experience with classes as large as are to be found in the laboratory of any school or college in the country. During the present year the author personally supervises the work of more than 180 different pupils in chemistry. This enables him not only to assure himself that the experiments of the book are practical, but that the directions for performing them are ample. It is found advisable to perform most of the experiments, with full explanation, in presence of the class, before requiring the pupils either to do the work or to recite the lesson. In the laboratory each pupil has a locker under his table, furnished with apparatus, as specified in the Appendix. Each has also the author's "Laboratory Manual," which contains on every left-hand page full directions for an experiment, with observations to be made, etc. The right-hand page is blank, and on that the pupil makes a record of his work. These notes are examined at the time, or subsequently, by the teacher, and the pupil is not allowed to take the book from the laboratory; nor can he use any other book on Chemistry while experimenting. By this means he learns to make his own observations and inferences.

For the benefit of the science and the added interest in the study, it is earnestly recommended that teachers encourage pupils to fit up laboratories of their own at home. This need not at first entail a large outlay. A small attic room with running water, a very few chemicals, and a little apparatus, are enough to begin with; these can be added to from time to time, as new material is wanted. In this way the student will find his love for science growing apace.

While endeavoring, by securing an able corps of critics, and in all other ways possible, to reduce errors to a minimum, the author disclaims any pretensions to a work entirely free from mistakes, holding himself alone responsible for any shortcomings, and trusting to the leniency of teachers and critics.

The manuscript has been read by Prof. Henry Carmichael, Ph.D., of Boston, and to his broad and accurate scholarship, as well as to his deep personal interest in the work, the author is indebted for much valuable and original matter. The following persons have generously read the proof, as a whole or in part, and made suggestions regarding it, and to them the author would return his thanks, as well as acknowledge his obligation: Prof. E. J. Bartlett, Dartmouth College, N.H.; Prof. F. C. Robinson, Bowdoin College, Me.; Prof. H. S. Carhart, Michigan University; Prof. B. D. Halsted, Iowa Agricultural College; Prof. W. T. Sedgwick, Institute of Technology, Boston; Pres. M. E. Wadsworth, Michigan Mining School; Prof. George Huntington, Carleton College, Minn.; Prof. Joseph Torrey, Iowa College; Mr. C. J. Lincoln, East Boston High School; Mr. W. H. Sylvester, English High School, Boston; Mr. F. W. Gilley, Chelsea, Mass., High School; the late D. S. Lewis, Chemist of the Boston Gas Works, and others.

R. P. W.

BOSTON, January 3, 1888.

TABLE OF CONTENTS.

CHAPTER I.
THE METRIC SYSTEM.

                                                                                  PAGE

Length. — Volume. — Weight 1

CHAPTER II.
DIVISIBILITY OF MATTER.

Mass. — Molecule. — Atom. — Element. — Compound. — Mixture. — Analysis. — Synthesis. — Metathesis. — Chemism 3

CHAPTER III.
MOLECULES AND ATOMS.

Synthesis 8

CHAPTER IV.
ELEMENTS AND BINARIES.

Symbols. — Names. — Coefficients. — Exponents. — Table of elements 10

CHAPTER V.
MANIPULATION.

To prepare and cut glass, etc. 14

CHAPTER VI.
OXYGEN.

Preparation. — Properties. — Combustion of carbon; sulphur; phosphorus; iron 17

CHAPTER VII.
NITROGEN.

Separation. — Properties 22

CHAPTER VIII.
HYDROGEN.

Preparation. — Properties. — Combustion. — Oxy-hydrogen blow-pipe 24

CHAPTER IX.
UNION BY WEIGHT.

Meaning of equations. — Problems 29

CHAPTER X.
CARBON.

Preparation. — Allotropic forms: diamond, graphite, amorphous carbon, coke, mineral coal. — Carbon a reducing agent, a decolorizer, disinfectant, absorber of gases 32

CHAPTER XI.
VALENCE.

Poles of attraction. — Radicals 38

CHAPTER XII.
ELECTRO-CHEMICAL RELATION OF ELEMENTS.

Deposition of silver; copper; lead. — Table of metals and non-metals, and discussion of their differences 41

CHAPTER XIII.
ELECTROLYSIS.

Decomposition of water and of salts. — Conclusions 44

CHAPTER XIV.
UNION BY VOLUME.

Avogadro's law and its applications 46

CHAPTER XV.
ACIDS AND BASES.

Characteristics of acids and bases. — Anhydrides. — Naming of acids. — Alkalies 49

CHAPTER XVI.
SALTS.

Preparation from acids and bases. — Naming of salts. — Occurrence 53

CHAPTER XVII.
CHLORHYDRIC ACID.

Preparation and tests. — Bromhydric, iodihydric, and fluorhydric acids. — Etching glass 56

CHAPTER XVIII.
NITRIC ACID.

Preparation, properties, tests, and uses. — Aqua regia: preparation and action 60

CHAPTER XIX.
SULPHURIC ACID.

Preparation, tests, manufacture, and importance. — Fuming sulphuric acid 63

CHAPTER XX.
AMMONIUM HYDRATE.

Preparation of bases. — Formation, preparation, tests, and uses of ammonia 67

CHAPTER XXI.
SODIUM HYDRATE.

Preparation and properties. — Potassium hydrate and calcium hydrate. 69

CHAPTER XXII.
OXIDES OF NITROGEN.

Nitrogen monoxide, dioxide, trioxide, tetroxide, pentoxide . 72

CHAPTER XXIII.
LAWS OF DEFINITE AND OF MULTIPLE PROPORTION,

and their application 75

CHAPTER XXIV.
CARBON PROTOXIDE

and water gas 77

CHAPTER XXV.
CARBON DIOXIDE.

Preparation and tests. — Oxidation in the human system. — Oxidation in water. — Deoxidation in plants 79

CHAPTER XXVI.
OZONE.

Description, preparation, and test 84

CHAPTER XXVII.
CHEMISTRY OF THE ATMOSPHERE.

Constituents of the air. — Air a mixture. — Water, carbon dioxide, and other ingredients of the atmosphere 86

CHAPTER XXVIII.
THE CHEMISTRY OF WATER.

Distillation of water. — Three states. — Pure water, sea-water, river-water, spring-water 88

CHAPTER XXIX.
THE CHEMISTRY OF FLAME.

Candle flame. — Bunsen flame. — Light and heat. — Temperature of combustion. — Oxidizing and reducing flames. — Combustible and supporter. — Explosive mixture of gases. — Generalizations 91

CHAPTER XXX.
CHLORINE.

Preparation. — Chlorine water. — Bleaching properties. — Disinfecting power. — A supporter of combustion. — Sources and uses 98

CHAPTER XXXI.
BROMINE.

Preparation. — Tests. — Description. — Uses 101

CHAPTER XXXII.
IODINE.

Preparation. — Tests. — Iodo-starch paper. — Occurrence. — Uses. — Fluorine 103

CHAPTER XXXIII.
THE HALOGENS.

Comparison. — Acids, oxides, and salts 106

CHAPTER XXXIV.
VAPOR DENSITY AND MOLECULAR WEIGHT.

Gaseous weights and volumes. — Vapor density defined. — Vapor density of oxygen 108

CHAPTER XXXV.
ATOMIC WEIGHT.

Definition. — Atomic weight of oxygen. — Molecular symbols. — Molecular and atomic volumes 111

CHAPTER XXXVI.

DIFFUSION AND CONDENSATION OF GASES.

Diffusion of gases. — Law of diffusion. — Cause. — Liquefaction and solidification of gases 114

CHAPTER XXXVII.

SULPHUR.

Separation. — Crystals from fusion. — Allotropy. — Solution. — Theory of Allotropy. — Occurrence and purification. — Uses. — Sulphur dioxide 116

CHAPTER XXXVIII.

HYDROGEN SULPHIDE.

Preparation. — Tests. — Combustion. — Uses. — An analyzer of metals. — Occurrence and properties 120

CHAPTER XXXIX.

PHOSPHORUS.

Solution and combustion. — Combustion under water. — Occurrence. — Sources. — Preparation of phosphates and phosphorus. — Properties. — Uses. — Matches. — Red phosphorus. — Phosphene 122

CHAPTER XL.

ARSENIC.

Separation. — Tests. — Expert analysis. — Properties and occurrence. — Atomic volume. — Uses of arsenic trioxide 126

CHAPTER XLI.

SILICON, SILICA, AND SILICATES.

Comparison of silicon and carbon. — Silica. — Silicates. — Formation of silica. 130

TABLE OF CONTENTS.

CHAPTER XLII.
GLASS AND POTTERY.

Glass an artificial silicate. — Manufacture. — Importance. — Porcelain and pottery. 132

CHAPTER XLIII.
METALS AND THEIR ALLOYS.

Comparison of metals and non-metals. — Alloys. — Low fusibility. — Amalgams. 135

CHAPTER XLIV.
SODIUM AND ITS COMPOUNDS.

Order of derivation. — Occurrence and preparation of sodium chloride; uses. — Sodium sulphate: manufacture and uses. — Sodium carbonate: occurrence, manufacture, and uses. — Sodium: preparation and uses. — Sodium hydrate: preparation and use. — Hydrogen sodium carbonate. — Sodium nitrate, 138

CHAPTER XLV.
POTASSIUM AND AMMONIUM.

Occurrence and preparation of potassium. — Potassium chlorate and cyanide. — Gunpowder. — Ammonium compounds . . . 143

CHAPTER XLVI.
CALCIUM COMPOUNDS.

Calcium carbonate. — Lime and its uses. — Hard water. — Formation of caves. — Calcium sulphate 146

CHAPTER XLVII.
MAGNESIUM, ALUMINIUM, AND ZINC.

Occurrence and preparation of magnesium. — Compounds of aluminium: reduction; properties, and uses. — Compounds, uses, and reduction of zinc 150

CHAPTER XLVIII.
IRON AND ITS COMPOUNDS.

Ores of iron. — Pig-iron. — Steel. — Wrought-iron. — Properties. — Salts of iron. — Change of valence and of color 154

CHAPTER XLIX.
LEAD AND TIN.

Distribution of lead. — Poisonous properties. — Some lead compounds. — Tin 161

CHAPTER L.
COPPER, MERCURY, AND SILVER.

Occurrence and uses of copper. — Compounds and uses of mercury. — Occurrence, reduction, and salts of silver 164

CHAPTER LI.
PHOTOGRAPHY.

Description 167

CHAPTER LII.
PLATINUM AND GOLD.

Methods of obtaining, and uses 169

CHAPTER LIII.
CHEMISTRY OF ROCKS.

Classification. — Composition. — Importance of siliceous rocks. — Soils. — Minerals. — The earth's interior. — Percentage of elements 171

CHAPTER LIV.
ORGANIC CHEMISTRY.

Comparison of organic and inorganic compounds. — Molecular differences. — Synthesis of organic compounds. — Marsh-gas

TABLE OF CONTENTS.   XV

PAGE

series. — Alcohols. — Ethers. — Other substitution products.
— Olefines and other series 174

CHAPTER LV.

ILLUMINATING GAS.

Source, preparation, purification, and composition. — Natural gas, 180

CHAPTER LVI.

ALCOHOL.

Fermented and distilled liquors. — Effect on the system. — Affinity for water. — Purity 184

CHAPTER LVII.

OILS, FATS, AND SOAPS.

Sources and kinds of oils and fats. — Saponification. — Manufacture and action of soap. — Glycerin, nitro-glycerin, and dynamite. — Butter and oleomargarine 186

CHAPTER LVIII.

CARBO-HYDRATES.

Sugars. — Glucose. — Starch. — Cellulose. — Gun-cotton. — Dextrin. — Zylonite 189

CHAPTER LIX.

CHEMISTRY OF FERMENTATION.

Ferments. — Alcoholic, acetic, and lactic fermentation. — Putrefaction. — Infectious diseases 193

CHAPTER LX.

CHEMISTRY OF LIFE.

Growth of minerals and of organic life. — Food of plants and of man. — Conservation of energy and of matter 196

CHAPTER LXI.

THEORIES.

The La Place theory. — Theory of evolution. — New theory of chemistry 199

CHAPTER LXII.

GAS VOLUMES AND WEIGHTS.

Quantitative experiments with oxygen and hydrogen. — Problems, 201

CHAPTER I.

THE METRIC SYSTEM.

1. The Metric System is the one here employed. A sufficient knowledge of it for use in the study of this book may be gained by means of the following experiments, which should be performed at the outset by each pupil.

2. Length.

Experiment 1. — Note the length of 10^{cm} (centimeters) on a metric ruler, as shown in Figure 1. Estimate by the eye alone this distance on the cover of a book, and then verify the result. Do the same on a t.t. (test-tube). Try this several times on different objects till you can carry in mind a tolerably accurate idea of 10^{cm}. About how many inches is it?

In the same way estimate the length of 1^{cm}, verifying each result. How does this compare with the distance between two blue lines of foolscap? Measure the diameter of the old nickel five-cent piece.

Next, try in the same way 5^{cm}. Carry each result in mind, taking such notes as may be necessary.

3. Capacity.

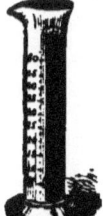

Fig. 2.

Experiment 2. — Into a graduate, shown in Figure 2, holding 25 or 50^{cc} (cubic centimeters) put 10^{cc} of water; then pour this into a t.t. Note, without marking, what proportion of the latter is filled; pour out the water, and again put into the t.t. the same quantity as nearly as can be estimated by the eye. Verify the re-

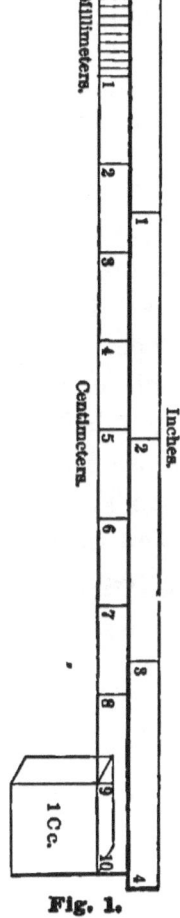

Fig. 1.

sult by pouring the water back into the graduate. Repeat several times until your estimate is quite accurate with a t.t. of given size. If you wish, try it with other sizes. Now estimate 1^{cc} of a liquid in a similar way. Do the same with 5^{cc}.

A cubic basin 10^{cm} on a side holds a liter. A liter contains $1,000^{cc}$. If filled with water, it weighs, under standard conditions, 1,000 grams. Verify by measurement.

4. Weight.

Experiment 3.—Put a small piece of paper on each pan of a pair of scales. On one place a 10^g (gram) weight. Balance this by placing fine salt on the other pan. Note the quantity as nearly as possible with the eye, then remove. Now put on the paper what you think is 10^g of salt. Verify by weighing. Repeat, as before, several times. Weigh 1^g, and estimate as before. Can 1^g of salt be piled on a one-cent coin? Experiment with 5^g.

5. Résumé.
—Lengths are measured in centimeters, liquids in cubic centimeters, solids in grams. In cases where it is not convenient to measure a liquid or weigh a solid, the estimates above will be near enough for most experiments herein given. Different solids of the same bulk of course differ in weight, but for one gram what can be piled on a one-cent piece may be called a sufficiently close estimate. The distance between two lines of foolscap is very nearly a centimeter. A cubic centimeter is seen in Figure 1. Temperatures are recorded in the centigrade scale.

CHAPTER II.

WHAT CHEMISTRY IS.

6. Divisibility of Matter.

Experiment 4.—Examine a few crystals of sugar, and crush them with the fingers. Grind them as fine as convenient, and examine with a lens. They are still capable of division. Put 3^g of sugar into a t.t., pour over it 5^{cc} of water, shake well, boil for a minute, holding the t.t. obliquely in the flame, using for the purpose a pair of wooden nippers (Fig. 3). If the sugar does not disappear, add more water. When cool, touch a drop of the liquid to the tongue. Evidently the sugar remains, though in a state too finely divided to be seen. This is called a *solution*, the sugar is said to be *soluble* in water, and water to be a *solvent* of sugar.

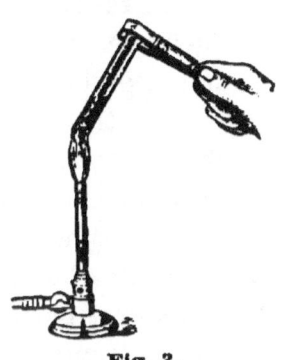

Fig. 3.

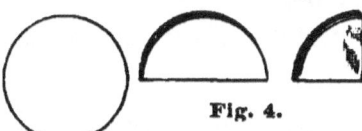

Fig. 4.

Fig. 5.

Now fold a filter paper, as in Figure 4, arrange it in a funnel (Fig. 5), and pour the solution upon it, catching what passes through, which is called the *filtrate*, in another t.t. that rests in a receiver (Fig. 5). After filtering, notice whether any *residue* is left on the filter paper. Taste a drop of the filtrate. Has sugar gone through the filter? If so, what do you infer of substances in solution passing through a filter? Save half the filtrate for Experiment 5, and dilute the other half with two or three times its own volume of water. Shake well, and taste.

We might have diluted the sugar solution many times more, and still the sweet taste would have remained. Thus the small quantity of sugar would be distributed through the whole mass, and be very finely divided.

By other experiments a much finer subdivision can be made. A solution of .00000002^g of the red coloring matter, fuchsine, in 1^{cc} of alcohol gives a distinct color.

Such experiments would seem to indicate that there is no limit to the divisibility of matter. But considerations which we cannot discuss here lead to the belief that such a limit does exist; that there are particles of sugar, and of all substances, which are incapable of further division without entirely changing the nature of the substance. To these smallest particles the name *molecules* is given.

A **mass** is any portion of a substance larger than a molecule; it is an aggregation of molecules.

? A **molecule** is the smallest particle of a substance that can exist alone.

<small>A substance in solution may be in a more finely divided state than otherwise, but it is not necessarily in its ultimate state of division.</small>

7. A Chemical Change. — Cannot this smallest particle of sugar, the molecule, be separated into still smaller particles of something else? May it not be a *compound* body, and will not some force separate it into two or more substances? The next experiment will answer the question.

<small>**Experiment 5.** — Take the sugar solution saved from Experiment 4, and add slowly 4^{cc} of strong sulphuric acid. Note any change of color, also the heat of the t.t. Add more acid if needed.</small>

A substance entirely different in color and properties has been formed. Now either the sugar, the acid, or the water has undergone a chemical change. It is, in fact, the sugar. But the molecule is the smallest particle of sugar possible. The acid must have either added something to the sugar molecules, or subtracted something from them. It was the latter. Here, then, is a force entirely different from the one which tends to reduce masses to molecules. The molecule has the same properties as the mass. Only a *physical* force was used in dissolving the sugar, and no heat was liberated. The acid has changed the sugar into a black mass, in fact into charcoal or carbon, and water; and heat has been produced. A *chemical* change has been brought about.

From this we see that molecules are not the ultimate divisions of matter. The smallest sugar particles are made up of still smaller particles of other things which do not resemble sugar, as a word is composed of letters which alone do not resemble the word. But can the charcoal itself be resolved into other substances, and these into still others, and so on? Carbon is one of the substances from which nothing else has been obtained. There are about seventy others which have not been resolved. These are called *elements*, and out of them are built all the compounds — mineral, vegetable, and animal — which we know.

8. An **element** is a chemically indivisible substance, or one from which nothing else can be extracted.

A **compound** is a substance which is made up of elements united in exact proportions by a force called chemism, or chemical affinity.

A **mixture** is composed of two or more elements or compounds blended together, but not held by any chemical attraction.

To which of these three classes does sugar belong? Carbon? The solution of sugar in water?

Carbon is an element; we call its smallest particle an atom.

An **atom** is the smallest particle of an element that can enter into combination. Atoms are indivisible and usually do not exist alone. Both elements and compounds have molecules, but only elements have atoms.

The molecule of an element usually contains two atoms; that of a compound may have two, or it may have hundreds. For a given compound the number is always definite.

Chemism is the force that binds atoms together to form molecules. The sugar molecule contains atoms, forty-five in all, of three different elements: carbon, hydrogen, and oxygen. That of salt has two atoms: one of sodium, one of chlorine. Should we say "an atom of sugar"? Why? Of what is a mass of sugar made up? A molecule? A mass of carbon? A molecule? Did the chemical affinity of the acid break up masses or molecules? In this respect it is a type of all chemical action. The distinction between physics and chemistry is here well shown. The molecule is the unit of the physicist, the atom that of the chemist. However large the masses changed by chemical action, that action is always on the individual molecule, the atoms of which are separated. If the molecule were an indivisible particle, no science of chemistry would be possible. The physicist finds the properties of masses of matter and resolves them into molecules, the chemist breaks up the molecule and from its atoms builds up other compounds.

Analysis is the separation of compounds into their elements.

Synthesis is the building up of compounds from their elements.

Of which is the sugar experiment an example?

Metathesis is an exchange of atoms in two different compounds; it gives rise to still other compounds.

A **chemical change** may add something to a substance, or subtract something from it, or it may both subtract and add, making a new substance with entirely different properties. Sulphur and carbon are two stable solids. The chemical union of the two forms a volatile liquid. A substance may be at one time a solid, at another a liquid, at another a gas, and yet not undergo any chemical change, because in each case the chemical composition is identical.

State which of these are chemical changes: rusting of iron, falling of rain, radiation of heat, souring of milk, evaporation of water, decay of vegetation, burning of wood, breaking of iron, bleaching of cloth. Give any other illustrations that occur to you.

Chemistry treats of matter in its simplest forms, and of the various combinations of those simplest forms.

CHAPTER III.

MOLECULES AND ATOMS.

9. Molecules are Extremely Small. — It has been estimated that a liter of any gas at 0° and 760^{mm} pressure contains 10^{24} molecules, *i.e.* one with twenty-four ciphers.

Thomson estimates that if a drop of water were magnified to the size of the earth, and its molecules increased in the same proportion, they would be larger than fine shot, but not so large as cricket balls.

A German has recently obtained a deposit of silver two-millionths of a millimeter thick, and visible to the naked eye. The computed diameter of the molecule is only one and a half millionths of a millimeter.

By a law of chemistry there is the same number of molecules in a given volume of every gas, if the temperature and pressure are the same. Hence, all *gaseous* molecules are of the same size, including, of course, the surrounding space. They are in rapid motion, and the lighter the gas the more rapid the motion. This gives rise to diffusion. See page 114.

10. We Know Nothing Definite of the Form of Molecules. — In this book they will always be represented as of the same size, that of two squares, ▭. A molecule is itself composed of atoms, — from two to several hundred. The size of the atom of most elements we represent by one square, ☐.

11. Atoms. — If the gaseous molecules be of the same size, it is clear that either the atoms themselves must be condensed, or the spaces between them must be smaller than before. We suppose the latter to be the case, and that they do not touch one another, the same thing being true of molecules. Atoms composing sugar must be crowded nearer together than those of salt. These atoms are probably in constant motion in the molecule, as the latter is in the mass. If we regard this square as a mass of matter, the dots may represent molecules; if we call it a molecule, the dots may be called atoms, though many molecules have no more than two or three atoms.

The following experiments illustrate the union of atoms to form molecules, and of elements to form compounds.

12. Union of Atoms.

Experiment 6. — Mix, on a paper, 5^g of iron turnings, and the same bulk of powdered sulphur, and transfer them to an ignition tube, a tube of hard glass for withstanding high temperatures. Hold the tube in the flame of a burner till the contents have become red-hot. After a minute break it by holding it under a jet of water. Put the contents into an evaporating-dish, and look for any uncombined iron or sulphur. Both iron and sulphur are elements. Is this an example of synthesis or of analysis? Why? Is the chemical union between masses of iron and sulphur, or between molecules, or between atoms? Is the product a compound, an element, or a mixture?

Experiment 7. — Try the same experiment, using copper instead of iron. The full explanation of these experiments is given on page 13.

CHAPTER IV.

ELEMENTS AND BINARIES.

13. About Seventy Different Elements are now Recognized, half of which have been discovered within little more than a century. These differ from one another in (1) atomic weight, (2) physical and chemical properties, (3) mode of occurrence, etc. Page 12 contains the most important elements.

The symbol of an element is usually the initial letter or letters of its Latin name, and stands for one atom of the element. C is the symbol for carbon, and represents one atom of it. O means one atom of oxygen.[1] Write, explain, and memorize the symbols of the elements in heavy type, page 12.

14. The Atomic Weight of an element is the weight of its atom compared with that of hydrogen. H is taken as the standard because it has the least atomic weight. The atomic weight of O is 16, which means that its atom weighs 16 times as much as the H atom. Every symbol, then, stands for a definite weight of the element, *i.e.* its atomic weight, as well as for its atom.

How much bromine by weight does Br stand for? What do these symbols mean — As, Na, N, P? If O represents one atom, how much does O_2 or 2 O stand for? How much by weight? Most elements have two

---

[1] The symbols of elements will also be used in this book to stand for an indefinite quantity of them; *e.g.* O will be used for oxygen in general as well as for one atom. The text will readily decide when symbols have a definite meaning, and when they are used in place of words.

atoms in the molecule. How many molecules in 6 H? 10 N? S_8? I_{20}?

The symbol of a compound is formed by writing in succession the symbols of the elements of which it is composed. How many atoms in the following molecules, and how many of each element: C_2H_6O? HNO_3? $PbSO_4$? $MgCl_2$? $Hg_2(NO_3)_2$?

15. The Simplest Compounds are Binaries. — A binary is a substance composed of two elements; *e.g.* common salt, which is a compound of sodium and chlorine. Its symbol is NaCl, its chemical name sodium chloride. The ending *ide* is applied to the last name of binaries. How many parts by weight of Na and of Cl in NaCl? What is the molecular weight, *i.e.* the weight of its molecule? Name KCl. How many atoms in its molecule? Parts by weight of each element? Molecular weight? Does the symbol stand for more than one molecule? How many molecules in 4 NaCl? How many atoms of Na and of Cl? Name these: HCl, NaBr, NaI, KBr, AgCl, AgI, HBr, HI, HF, HgO, ZnO, ZnS, MgO, CaO. Compute the proportion by weight of each element in the last three.

A **coefficient** before the symbol of a compound includes all the elements of the symbol, and shows the number of molecules. How many in these: 6 KBr? 3 SnO? 12 NaCl? How many atoms of each element in the above?

An **exponent**, always written below, applies only to the element after which it is written, and shows the number of atoms. Explain these: $AuCl_3$, $ZnCl_2$, Hg_2Cl_2.

Write symbols for four molecules of sodium bromide, one of silver iodide (always omit coefficient one), eight of potassium bromide, ten of hydrogen chloride; also for one molecule of each of these: hydrogen fluoride, potassium iodide, silver chloride.

ELEMENTS AND BINARIES.

In all the above cases the elements have united atom for atom. Some elements will not so unite. In $CaCl_2$ how many atoms of each element? Parts by weight of each? Give molecular weight. Is the size of the molecule thereby changed? See page 8. Name these, give the number of atoms of each element in the molecule, and the proportion by weight, also their molecular weights: $AuCl_3$, $ZnCl_2$, $MnCl_2$, Na_2O, K_2S, H_3P, H_4C.

Principal Elements.

Name.	Sym.	At. Wt.	Valence.	Vap. D.	At. Vol.	Mol. Vol.	State.
Aluminium	Al	27.	II, IV	...	...	...	Solid
Antimony	Sb	120.	III, V	...	...	...	"
Arsenic	As	75.	III, V	150.	□	⬜	"
Barium	Ba	137.	II	...	...	...	"
Bismuth	Bi	210.	III, V	...	...	...	"
Boron	B	11.	III	...	...	...	"
Bromine	Br	80.	I, (V)	80.	□	⬜	Liquid
Cadmium	Cd	112.	II	56.	⬜	⬜	Solid
Calcium	Ca	40.	II	...	...	...	"
Carbon	C	12.	(II), IV	...	...	...	"
Chlorine	Cl	35.5	I, (V)	35.5	□	⬜	Gas
Chromium	Cr	52.	(II), IV, VI	...	...	...	Solid
Cobalt	Co	59.	II, IV	...	...	...	"
Copper	Cu	63.	I, II	...	...	...	"
Fluorine	F	19.	I, (V)	...	...	...	Gas
Gold	Au	196.	(I), III	...	...	...	Solid
Hydrogen	H	1.	I	1.	□	⬜	Gas
Iodine	I	127.	I, (V)	127.	...	...	Solid
Iron	Fe	56.	II, IV, (VI)	...	...	...	"
Lead	Pb	206.	II, IV	...	...	...	"
Lithium	Li	7.	I	...	...	...	"
Magnesium	Mg	24.	II	...	...	...	"
Manganese	Mn	55.	II, IV, VI	...	...	...	"
Mercury	Hg	200.	I, II	100.	⬜	⬜	Liquid
Nickel	Ni	59.	II, IV	...	...	...	Solid
Nitrogen	N	14.	(I), III, V	14.	□	⬜	Gas
Oxygen	O	16.	II	16.	□	⬜	"
Phosphorus	P	31.	(I), III, V	62.	□	⬜	Solid
Platinum	Pt	197.	(II), IV	...	...	...	"
Potassium	K	39.	I	...	...	...	"
Silicon	Si	28.	IV	...	...	...	"
Silver	Ag	108.	I	...	...	...	"
Sodium	Na	23.	I	...	...	...	"
Strontium	Sr	87.	II	...	...	...	"
Sulphur	S	32.	II, IV, (VI)	32 (96)	□	⬜	"
Tin	Sn	118.	II, IV	...	...	...	"
Zinc	Zn	65.	II	32.5	⬜	⬜	"

ELEMENTS AND BINARIES. 13

If more than one atom of an element enters into the composition of a binary, a prefix is often used to denote the number. SO_2 is called sulphur dioxide, to distinguish it from SO_3, sulphur trioxide. Name these: CO_2, SiO_2, MnO_2. The prefixes are: *mono* or *proto*, one; *di* or *bi*, two; *tri* or *ter*, three; *tetra*, four; *pente*, five; *hex*, six; etc. Diarsenic pentoxide is written, As_2O_5. Symbolize these: carbon protoxide, diphosphorus pentoxide, diphosphorus trioxide, iron disulphide, iron protosulphide. Often only the prefix of the last name is used.

16. An Oxide is a Compound of Oxygen and Some Other Element, as HgO. What is a chloride? Define sulphide, phosphide, arsenide, carbide, bromide, iodide, fluoride.

In Experiment 6, where S and Fe united, the symbol of the product was FeS. Name it. How many parts by weight of each element? What is its molecular weight? To produce FeS a chemical union took place between each atom of the Fe and of the S. We may express this reaction, *i.e.* chemical action, by an equation: —

$$\text{Iron} + \text{Sulphur} = \text{Iron sulphide.}$$
Or, using symbols, $Fe + S = FeS.$
Using atomic weights, $56 + 32 = 88.$

These equations are explained by saying that 56 parts by weight of iron unite chemically with 32 parts by weight of sulphur to produce 88 parts by weight of iron sulphide. This, then, indicates the proportion of each element which combines, and which should be taken for the experiment. If 56^g of Fe be used, 32^g of S should be taken. If we use more than 56 parts of Fe with 32 of S, will it all combine? If more than 32 of S with 56 of Fe? There is found to be a definite quantity of each element in every chemical compound. Symbols would have no meaning if this were not so.

Write and explain the equation for the experiment with copper and sulphur, using names, symbols, and weights, as above.

CHAPTER V.

MANIPULATION.

17. To Break Glass Tubing.

Experiment 8. — Lay the tubing on a flat surface, and draw a sharp three-cornered file two or three times at right angles across it where it is to be broken, till a scratch is made. Take the tube in the hands, having the two thumbs nearly opposite the scratch, and the fingers on the other side. Press outward quickly with the thumbs, and at the same time pull the hands strongly apart, and the tubing should break squarely at the scratch.

To break large tubing, or cut off bottles, lamp chimneys, etc., first make a scratch as before; then heat the handle of a file, or a blunt iron — in a blast-lamp flame by preference — till it is red-hot, and at once press it against the scratch till the glass begins to crack. The fracture can be led in any direction by keeping the iron just in front of it. Re-heat the iron as often as necessary.

18. To Make Ignition-Tubes.

Experiment 9. — Hold the glass tubing between the thumb and forefinger of each hand, resting it against the second finger. Heat it in the upper flame, slowly at first, then strongly, but heat only a very small portion in length, and keep it in constant rotation with the right hand. Hold it steadily, and avoid twisting it as the glass softens. The yielding is detected by the yellow flame above the glass and by an uneven pressure on the hands. Pull it a little as it yields, then heat a part just at one side of the most softened portion. Rotate constantly without twisting, and soon it can be separated into two closed tubes. No thread should be attached; but if there be one, it can be broken off and the end welded. The bottom can be made more symmetrical by heating it red-hot, then blowing, gradually, into the open end, this being inserted in the mouth. The parts should be annealed by holding above the flame for a short time, to cool slowly.

For hard glass — Bohemian — or large tubes, the blast-lamp or blowpipe is needed. In the blast-lamp air is forced out with illuminating gas. This gives a high degree of heat. Bulbs can be made in the same way as ignition-tubes, and thistle-tubes are made by blowing out the end of a heated bulb, and rounding it with charcoal.

19. To Bend Glass Tubing.

Experiment 10. — Hold the tube in the upper flame. Rotate it so as to heat all parts equally, and let the flame spread over 3 or 4^{cm} in length. When the glass begins to yield, without removing from the flame slowly bend it as desired. Avoid twisting, and be sure to have all parts in the same plane; also avoid bending too quickly, if you would have a well-rounded joint. Anneal each bend as made. Heated glass of any kind should never be brought in contact with a cool body. For making O, H, etc., a glass tube

Fig. 6.

— delivery-tube — 50^{cm} long should have three bends, as in Figure 6. The pupil should first experiment with short pieces of glass, 10 or 15^{cm} long. An ordinary gas flame is the best for bending glass.

20. To Cut Glass.

Experiment 11. — Lay the glass plate on a flat surface, and draw a steel glass-cutter — revolving wheel — over it, holding this against a ruler for a guide, and pressing down hard enough to scratch the glass. Then break it by holding between the thumb and fingers, having the thumbs on the side opposite to the scratch, and pressing them outward while bending the ends of the glass inward. The break will follow the scratch.

Holes can be bored through glass and bottles with a broken end of a round file kept wet with a solution of camphor in oil of turpentine.

21. To Perforate Corks.

Experiment 12. — First make a small hole in the cork with the pointed handle of a round — rat-tail — file. Have the hole perpendicular to the surface of the cork. This can be done by holding the cork in the left hand and pressing against the larger surface, or upper part, of the cork, with the file in the right hand. Only a mere opening is made in this way, which must be enlarged by the other end of the file. A second or third file of larger size may be employed, according to the size of the hole to be made, which must be a little smaller than the tube it is to receive, and perfectly round.

CHAPTER VI.

OXYGEN.

22. To Obtain Oxygen.

Experiment 13. — Take 5^g of crystals of potassium chlorate ($KClO_3$) and, without pulverizing, mix with the same weight of pure powdered manganese dioxide (MnO_2). Put the mixture into a t.t., and insert a d.t. — delivery-tube — having the cork fit tightly. Hang it on a r.s. — ring-stand, — as in Figure 7, having the other end of the d.t.

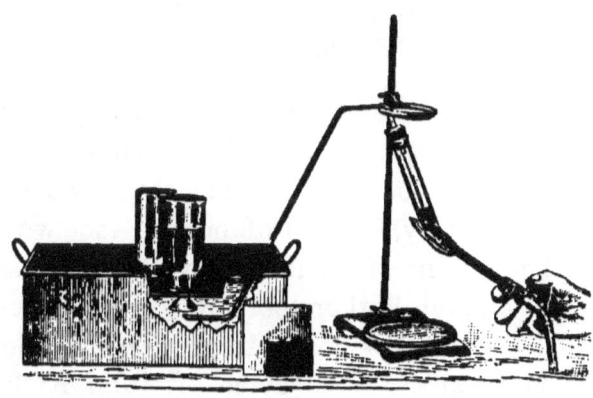

Fig. 7.

under the shelf, in a pneumatic trough, filled with water just above the shelf. Fill three or more receivers — wide-mouthed bottles — with water, cover the mouth of each with a glass plate, invert it with its mouth under water, and put it on the shelf of the trough, removing the plate. No air should be in the bottles. Have the end of the d.t. so that the gas will rise through the orifice. Hold a lighted lamp in the hand, and bring the flame against the mixture in the t.t. Keep the lamp slightly in motion, with the hand, so as not to break the t.t. by over-heating in one place. Heat the mixture strongly, if necessary. The upper part of the t.t. it filled with air: allow this to escape for a few seconds; then move a receiver over the orifice, and fill it with gas. As soon as the lamp is taken away, remove the d.t. from the water. The gas contracts, on cooling, and if not removed, water will be drawn over, and the t.t. will be broken. Let the t.t. hang on the r.s. till cool.

With glass plates take out the receivers, leaving them covered, mouth upward (Fig. 8), with little or no water inside. When cool, the t.t. may be cleaned with water, by covering its mouth with the thumb or hand, and shaking it vigorously.

Fig. 8.

What elements, and how many, in $KClO_3$? In MnO_2? It is evident that each of these compounds contains O. Why, then, could we not have taken either separately, instead of mixing the two? This could have been done at a sufficiently high temperature. MnO_2 requires a much higher temperature for dissociation, *i.e.* separation into its elements, than $KClO_3$, while a mixture of the two causes O to come off from $KClO_3$ at a lower temperature than if alone. It is not known that MnO_2 suffers any change.

Each molecule of potassium chlorate undergoes the following change: —

$$\text{Potassium Chlorate} = \text{Potassium Chloride} + \text{Oxygen}.$$
$$KClO_3 = KCl + 3O.$$

Is this analysis or synthesis? Complete the equation, by using weights, and explain it. Notice whether the right-hand member of the equation has the same number of atoms as the left. Has anything been lost or gained? What element has heat separated? Does the experiment show whether O is very soluble in water? How many grams of O are obtainable from 122.5^g $KClO_3$?

OXYGEN.

PROPERTIES.

23. Combustion of Carbon.

Experiment 14. — Examine the gas in one of the receivers. Put a lighted splinter into the receiver, sliding along the glass cover. Remove it, blow it out, and put in again while glowing. Is it rekindled? Repeat till it will no longer burn. Is the gas a supporter of combustion? How did the combustion compare with that in air? Is it probable that air is pure O? Why did the flame at last go out? Has the O been destroyed, or chemically united with something else?

Wood is in part C. CO_2 is formed by the combustion; name it. The equation is $C + 2O = CO_2$. Affix the names and weights. Is CO_2 a supporter of combustion? Note that when C is burned with plenty of O, CO_2 is always formed, and that no matter how great the conflagration, the union is atom by atom. Combustion, as here shown, is only a rapid union of O with some other substance, as C or H.

24. Combustion of Sulphur.

Experiment 15. — Hollow out one end of a piece of electric-light pencil, or of crayon, 3^{cm} long, and attach it to a Cu wire (Fig. 9). Put into this a piece of S as large as a pea, ignite it by holding in the flame, and then hold it in a receiver of O. Note the color and brightness of the flame, and compare with the same in the air. Also note the color and odor of the product. The new gas is SO_2. Name it, and write the equation for its production from S and O. How do you almost daily perform a similar experiment? Is the product a supporter of combustion?

Fig. 9.

25. Combustion of Phosphorus.

Experiment 16. — With forceps, which should always be used in handling this element, put a bit of P, half as large as the S above, into the crayon, called a deflagrating-spoon. Heat another wire, touch it to the P, and at once lower the latter into a receiver of O. Notice the combustion, the color of the flame and of the product. After removing, be sure to burn every bit of P by holding it in a flame, as it is liable to take fire if left. The product of the combustion is a union of what two elements? Is it an oxide? Its symbol is P_2O_5. Write the equation, using symbols, names, and weights. Towards the close of the experiment, when the O is nearly all combined, P_2O_3 is formed, as it is also when P oxidizes at a low temperature. Name it and write the equation.

26. Combustion of Iron.

Experiment 17. — Take in the forceps a piece of iron picture-cord wire 6 or 8^{cm} long, hold one end in the flame for an instant, then dip it into some S. Enough S will adhere to be set on fire by holding it in the flame again. Then at once dip it into a receiver of O with a little water in the bottom. The iron will burn with scintillations. Is this analysis or synthesis? What elements combine? A watch-spring, heated to take out the temper, may be used, but picture-wire is better.

The product is Fe_3O_4. Write the equation. How much Fe by weight in the formula? How much O? What *per cent* by weight of Fe in the compound? Multiply the fractional part by 100. What per cent of O? What per cent of CO_2 is C? O_2? Find the percentage composition of SO_2. P_2O_5.

From the last five experiments what do you infer of the tendency of O to unite with other elements?

27. Oxygen is a Gas without Color, Odor, or Taste.
— It is chemically a very active element; that is, it unites with almost everything. Fluorine is the only element with which it will not combine. When oxygen combines with a single element, what is the compound called? We have found that O makes up a certain portion of the air; later, we shall see how large the proportion is. Its tendency to combine with almost everything is a reason for the decay, rust, and oxidation of so many substances, and for conflagrations, great and small. New compounds are thus

formed, of which O constitutes one factor. Water, H_2O, is only a chemical union of O and H. Iron rust, Fe_2O_3 and H_2O, is composed of O, Fe, and water. The burning of wood or of coal gives rise to carbon dioxide, CO_2, and water. Decay of animal and vegetable matter is hastened by this all-pervading element. O forms a portion of all animal and vegetable matter, of almost all rocks and minerals, and of water. It is the most abundant of all elements, and makes up from one-half to two-thirds of the earth's surface. Compute the proportion of it, by weight, in water, H_2O. It is the union of O in the air with C and H in our blood that keeps up the heat of the body and supports life. See page 81.

There are many ways of preparing this element besides the one given above. It may be obtained from water (Experiment 38) and from many other compounds, *e.g.* by heating mercury oxide, HgO.

CHAPTER VII.

NITROGEN.

28. Separation.

Experiment 18. — Fasten a piece of electric-light pencil, or of crayon, to a wire, as in Experiment 15, and bend the wire so it will reach half-way to the bottom of a receiver. Using forceps, put into the crayon a small piece of phosphorus. Pass the wire up through the orifice in the shelf of a p.t. (pneumatic trough), having water at least 1^{cm} above the shelf. Heat another wire, touch it to the P, and quickly invert an empty receiver over the P, having the mouth under water, so as to admit no air (Fig. 10). Let the P burn as long as it will, then remove the wire and the crayon, letting in no air. Note the color of the product, and leave till it is tolerably clear, then remove the receiver with a glass plate, leaving the water in the bottom.

Fig. 10.

Do the fumes resemble those of Experiment 16? Does it seem likely that part of the air is O? Why a part only? Find what proportion of the receiver is filled with water by measuring the water with a graduate; then fill it with water and measure that; compute the percentage which the former is of the latter. What proportion of the air, then, is O? What was the only means of escape for the P_2O_5 and P_2O_3 formed? These products are solids. Are they soluble in water? Compute the percentage composition, always by weight, of P_2O_3 and P_2O_5.

The gas left in the receiver is evidently not O. Experiment 19 will prove this conclusively, and show the properties of the new gas.

29. Properties.

Experiment 19. — When the white cloud has disappeared, slide the plate along, and insert a burning stick; try one that still glows.

See whether the P and S on the end of a match will burn. Is the gas a supporter of combustion? Since it does not unite with C, S, or P, is it an active or a passive element? Compare it with O. Air is about 14½ times as heavy as H. Which is heavier, air or N? See page 12. Air or O?

Write out the chief properties, physical and chemical, of N, as found in this experiment.

30. Inactivity of N. — N will scarcely unite chemically except on being set free from compounds. It has, however, an intense affinity for boron, and will even go through a carbon crucible to unite with it. It is not combined with O in the air; but the two form a mixture (page 86), of which N makes up four-fifths, its use being to dilute the O. What would be the effect, in case of a fire, if air were pure O? What effect on the human system?

Growing plants need a great deal of N, but they are incapable of making use of that in the air, on account of the chemical inactivity of the element. Their supply comes from compounds in earth, water, and air. By reason of its inertness N is very easily set free from its compounds. For this reason it is a constituent of most explosives, as gunpowder, nitro-glycerine, dynamite, etc. These solids, by heat or concussion, are suddenly changed to gases, which thereby occupy much more space, causing an explosion.

Nitrogen exists in many compounds, such as the nitrates; but the great source of it all is the atmosphere. See page 85.

CHAPTER VIII.

HYDROGEN.

31. Preparation.

Experiment 20. — Prepare apparatus as for making O. Be sure that the cork perfectly fits both d.t. and t.t., or the H will escape.

Fig. 11.

Cover 5^g granulated Zn, in the t.t., with 10^{cc} H$_2$O, and add 5^{cc} chlorhydric acid, HCl. Adjust as for O (Fig. 7), except that no heat is to be applied. If the action is not brisk enough, add more HCl. Collect several receivers of the gas over water, adding small quantities of HCl when necessary. Observe the black floating residuum; it is carbon, lead, etc. With a glass plate remove the receivers, keeping them inverted (Fig. 11), or the H will escape.

32. The Chemical Change is as follows: —

$$\text{Zinc} + \text{hydrogen chloride} = \text{zinc chloride} + \text{hydrogen.}$$
$$\text{Zn} + 2\,\text{HCl} = \text{ZnCl}_2 + 2\,\text{H.}$$

Complete by adding the weights, and explain. Notice that the water does not take part in the change; it is added to dissolve the ZnCl$_2$ formed, and thus keep it from coating the Zn and preventing further action of the acid. Note also that Zn has simply changed places with H, one atom of the former having driven off two atoms of the latter. The H, having nothing to unite with, is set free as a gas, and collected over water. Of course Zn must have a stronger chemical affinity for Cl than H has, or the change could not have taken place. Why one Zn atom replaces two H atoms will be explained later, as far as an explanation is possible. This equation should be studied carefully, as a type of all equations. The left-hand member shows what were taken, *i.e.* the factors; the right-hand shows what were obtained, *i.e.* the products. H_2SO_4 might have been used instead of HCl. In that case the reaction, or equation, would have been: —

$$\text{Zinc} + \text{hydrogen sulphate} = \text{zinc sulphate} + \text{hydrogen}.$$
$$Zn + H_2SO_4 = ZnSO_4 + 2H.$$

Iron might have been used instead of zinc, in which case the reactions would have been: —

$$\text{Iron} + \text{hydrogen chloride} = \text{iron chloride} + \text{hydrogen}.$$
$$Fe + 2HCl = FeCl_2 + 2H.$$

$$\text{Iron} + \text{hydrogen sulphate} = \text{iron sulphate} + \text{hydrogen}.$$
$$Fe + H_2SO_4 = FeSO_4 + 2H.$$

Write the weights and explain the equations. The latter should be memorized.

33. Properties.

Experiment 21. — Lift with the left hand a receiver of H, still inverted, and insert a burning splinter with the right (Fig. 12). Does the splinter continue to burn? Does the gas burn? If so, where? Is the light brilliant? Note the color of the flame. Is there any explosion? Try this experiment with several receivers. Is the gas a supporter of combustion? *i.e.* will carbon burn in it? Is it combustible? *i.e.* does it burn? If so, it unites with some part of the air. With what part?

Fig. 12.

34. Collecting H by Upward Displacement.

Experiment 22. — Pass a d.t. from a H generator to the top of a receiver or t.t. (Fig. 13).

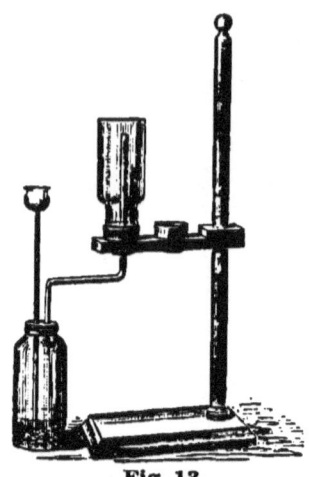

Fig. 13.

The escaping H being so much lighter than air will force the latter down. To obtain the gas unmixed with air, the d.t. should tightly fit a cardboard placed under the mouth of the receiver. When filled, the receiver can be removed, inverted as usual, and the gas tested. In this and other experiments for generating H, a thistle-tube, the end of which dips under the liquid, can be used for pouring in acid, as in Figure 13.

35. Philosopher's Lamp and Musical Flame.

Experiment 23. — Fit to a cork a piece of glass tubing 10 or 15^{cm} long, having the outer end drawn out to a point with a small opening, and insert it in the H generator. Before igniting the gas at the end of the tube take the precaution to collect a t.t. of it by upward displacement, and bring this in contact with a flame. If a sharp explosion ensues, air is not wholly expelled from the generator, and it would be dangerous to light the gas. When no sound, or very little, follows, light the escaping gas. The generation of H must not be too rapid, neither should the t.t. be held under the face, as the cork is liable to be forced out by the pressure of H. A safety-tube, similar to the thistle-tube above, will prevent this. This apparatus is called the "philosopher's lamp." Thrust the flame into a long glass tube 1½ to 3^{cm} in diameter, as shown in Figure 14, and listen for a musical note.

Fig. 14.

36. Product of Burning H in Air.

Experiment 24. — Fill a tube 2 or 3^{cm} in diameter with calcium chloride, $CaCl_2$, and connect one end with a generator of H (Fig. 15). At the other end have a philosopher's lamp-tube.

Observing the usual precautions, light the gas and hold over it a receiver, till quite a quantity of moisture collects. All water was taken from the gas by the dryer, $CaCl_2$. What is, therefore, the product of burning H in air? Complete this equation and explain it: $2H + O = ?$ Figure 16 shows a drying apparatus arranged to hold $CaCl_2$.

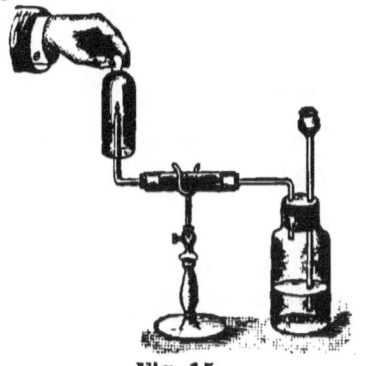
Fig. 15.

Fig. 16.

37. Explosiveness of H.

Experiment 25. — Fill a soda-water bottle of thick glass with water, invert it in a pneumatic trough, and collect not over ¼ full of H. Now remove the bottle, still inverted, letting air in to fill the other ¾. Mix the air and H by covering the mouth of the bottle with the hand, and shaking well; then hold the mouth of the bottle, slightly inclined, in a flame. Explain the explosion which follows. If ¾ was air, what part was O? What use did the N serve? Note any danger in exploding H mixed with pure O. What proportions of O and H by volume would be most dangerously explosive? See page 46. What proportion by weight?

By the rapid union of the two elements, the high temperature suddenly expanded the gaseous product, which immediately contracted; both expansion and contraction produced the noise of explosion.

38. Pure H is a Gas without Color, Odor, or Taste.
— It is the lightest of the elements, $14\frac{1}{2}$ times as light as

air. It occurs uncombined in coal-mines, and some other places, but the readiness with which it unites with other elements, particularly O, prevents its accumulation in large quantities. It constitutes two-thirds of the volume of the gases resulting from the decomposition of water, and one-ninth of the weight. Compute the latter from its symbol. It is a constituent of plants and animals, and some rocks. Considering the volume of the ocean, the total amount of H is large. It can be separated from H₂O by electrolysis (page 44), or by C, as in the manufacture of water gas (page 78).

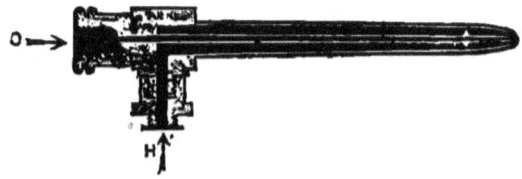

Fig. 17.

When burned with O it forms H₂O. Pure O and H when burning give great heat, but little light. The oxyhydrogen blow-pipe (Fig. 17) is a device for producing the highest temperatures of combustion. It has O in the inner tube and H in the outer. Why would it not be better the other way? These unite at the end, and are burned, giving great heat. A piece of lime put into the flame gives the brilliant Drummond or calcium light.

CHAPTER IX.

UNION BY WEIGHT.

39. In the Equation —

$$Zn + 2\,HCl = ZnCl_2 + 2\,H$$
$$65 + 73 = 136 + 2$$

65 parts by weight of Zn are required to liberate 2 parts by weight of H; or, by using 65g Zn with 73g HCl, we obtain 2g H. If twice as much Zn (130g) were used, 4g H could be obtained, with, of course, twice as much HCl. With 260g Zn, how much H could be liberated? A proportion may be made as follows: —

Zn given : Zn required : : H given : H required.[1]
65 : 260 : : 2 : x.

Solving, we have 8g H.

How much H is obtainable by using 5g Zn, as in the experiment?

To avoid error in solving similar problems, the best plan is as follows: —

$Zn + 2\,HCl = ZnCl_2 + 2\,H$	$65 : 5 : : 2 : x$
65 2	$65x = 10$
5 x	$x = \tfrac{10}{65} = \tfrac{2}{13}$ *Ans.* $\tfrac{2}{13}$g.

The equation should first be written; next, the atomic or molecular weights which you wish to use, and only those, to avoid confusion; then, on the third line, the quantity of the substance to be used, with x underneath the substance wanted. The example above will best show this. This plan will prevent the possibility of error. The proportion will then be: —

a given : a required : : b given : b required.

How much Zn is required to produce 30g H?

---

[1] Given, as here used, means the weight called for by the equation; required means that called for by the question.

UNION BY WEIGHT.

$$Zn + 2\,HCl = ZnCl_2 + 2\,H \qquad 2:30::65:x$$
$$65 \qquad\qquad\qquad 2 \qquad\qquad 2x = 1950$$
$$x \qquad\qquad\qquad 30 \qquad\qquad x = 975 \qquad Ans.\ 975^g\ Zn.$$

Solve: —

(1) How much Zn is necessary for 14^g H?
(2) How many pounds of Zn are necessary for 3 pounds of H?
(3) How many grams of H from 17^g of Zn?
(4) How many tons of H from ½ ton of Zn?

Suppose we wish to find how much chlorhydric acid — pure gas — will give 12^g H. The question involves only HCl and H. Arrange as follows: —

$$Zn + 2\,HCl = ZnCl_2 + 2\,H \qquad H\ giv.:H\ req.::HCl\ giv.:HCl\ req.$$
$$73 \qquad\qquad 2 \qquad\qquad 2\ :\ 12\ ::\ 73\ :\ x$$
$$x \qquad\qquad 12 \qquad\qquad 2x = 876 \qquad x = 438$$
$$Ans.\ 438^g\ HCl.$$

Solve: —

(1) How much HCl is needed to produce 100^g H?
(2) How much H in 10^g HCl?
(3) How much $ZnCl_2$ is formed by using 50^g HCl? The question is now between HCl and $ZnCl_2$.

$$Zn + 2\,HCl = ZnCl_2 + 2\,H$$
$$73 \qquad 136 \qquad\qquad \text{Arrange the proportion, and solve.}$$
$$50 \qquad x$$

Suppose we have generated H by using H_2SO_4: the equation is $Zn + H_2SO_4 = ZnSO_4 + 2\,H$. There is the same relation as before between the quantities of Zn and of H, but the H_2SO_4 and $ZnSO_4$ are different.

How much H_2SO_4 is needed to generate 12^g H?

$$Zn + H_2SO_4 = ZnSO_4 + 2\,H$$
$$98 \qquad\qquad 2 \qquad \text{Make the proportion, and solve.}$$
$$x \qquad\qquad 12$$

Solve: —

(1) How much H in 200^g H_2SO_4?
(2) How much $ZnSO_4$ is produced from 200^g H_2SO_4?

(3) How much H_2SO_4 is needed for $7\frac{1}{2}$g H?
(4) How much Zn will 40g H_2SO_4 combine with?
(5) How much Fe will 40g H_2SO_4 combine with? See page 25.
(6) How much H can be obtained by using 75g Fe?

These principles apply to all reactions. Suppose, for example, we wish to get 10g of O: how much $KClO_3$ will it be necessary to use? The reaction is:—

$KClO_3 = KCl + O_3$
122.5   48
x 10

48 : 10 :: 122.5 : x

Ans. 25.5+g $KClO_3$.

The pupil should be required to make up problems of his own, using various reactions, and to solve them.

CHAPTER X.

CARBON.

Examine graphite, anthracite coal, bituminous coal, cannel coal, wood, gas carbon, coke.

40. Preparation of C.

Experiment 26. — Hold a porcelain dish or a plate in the flame of a candle, or of a Bunsen burner with the openings at the bottom closed. After a minute examine the deposit. It is carbon, *i.e.* lamp-black or soot, which is a constituent of gas, or of the candle. Open the valve at the base of the Bunsen burner, and hold the deposit in the flame. Does the C gradually disappear? If so, it has been burned to CO_2. $C + 2O = CO_2$. Is C a combustible element?

Experiment 27. — Ignite a splinter, and observe the combustion and the smoke, if any. Try to collect some C in the same way as before.

With plenty of O and high enough temperature, all the C is burned to CO_2, whether in gas, candle, or wood. CO_2 is an invisible gas. The porcelain, when held in the flame, cools the C below the point at which it burns, called the kindling-point, and hence it is deposited. The greater part of smoke is unburned carbon.

Experiment 28. — Hold an inverted dry t.t. or receiver over the flame of a burning candle, and look for any moisture (H_2O). What two elements are shown by these experiments to exist in the candle? The same two are found in wood and in gas.

Experiment 29. — Put into a small Hessian crucible (Fig. 18) some pieces of wood 2 or 3^{cm} long, cover with sand, and heat the crucible strongly. When smoking stops, cool the crucible, remove the contents, and examine the charcoal. The gases have been driven off from the wood, and the greater part of what is left is C.

Experiment 30. — Put 1^g of sugar into a porcelain crucible, and heat till the sugar is black. C is left. See Experiment 5. Remove the C with a strong solution of sodium hydrate (page 208).

Fig. 18.

41. Allotropic Forms. — Carbon is peculiar in that it occurs in at least three allotropic, *i.e.* different, forms, all having different properties. These are diamond, graphite, and amorphous — not crystalline — carbon. The latter includes charcoal, lamp-black, bone-black, gas carbon, coke, and mineral coal. All these forms of C have one property in common; they burn in O at a high temperature, forming CO_2. This proves that each is the element C, though it is often mixed with some impurities.

Allotropy, or allotropism, is the quality which an element often has of appearing under various forms, with different properties. The forms of C are a good illustration.

42. Diamond is the purest C; but even this in burning leaves a little ash, showing that it is not quite pure. It is a rare mineral, found in India, South Africa, and Brazil, and is the hardest and most highly refractive to light of all minerals. Boron is harder.[1] When heated in the electric arc, at very high temperatures, diamond swells and turns black.

---

[1] B, not occurring free, is not a mineral.

43. Graphite, or Plumbago, is One of the Softest Minerals. — It is black and infusible, and oxidizes only at very high temperatures, higher than the diamond. It contains from 95 to 98 per cent C. Graphite is found in the oldest rock formations, in the United States and Siberia. It is artificially formed in the iron furnace. Graphite is employed for crucibles where great heat is required, for a lubricant, for making metal castings, and, mixed with clay, for lead-pencils. It is often called black-lead.

44. Amorphous Carbon comprises the following varieties.

Charcoal is made by heating wood, for a long time, out of contact with the air. The volatile gases are thus driven off from the wood; what is left is C, and a small quantity of mineral matter which remains as ash when the coal is burned.

45. Lamp-black is prepared as in Experiment 26, or by igniting turpentine ($C_{10}H_{16}$), naphtha, and various oils, and collecting the C of the smoke. It is used for making printers' ink, India ink, etc. A very pure variety is obtained from natural gas.

Bone-black, or animal charcoal, is obtained by distilling bones, *i.e.* by heating them in retorts into which no air is admitted. The C is the charred residue.

Gas Carbon is formed in the retorts of the gas-house. See page 182. It is used to some extent in electrical work.

46. Coke is the residue left after distilling soft coal. It is tolerably pure carbon, with some ash and a little volatile matter. It burns without flame.

47. Mineral Coal is fossilized wood or other vegetable matter. Millions of years ago trees and other vegetation covered the earth as they do to-day. In certain places they slowly sank, together with the land, into the interior of the earth, were covered with sand, rock, and water, and heated from the earth's interior. A slow distillation took place, which drove off some of the gases, and converted vegetable matter into coal. All the coal dug from the earth represents vegetable life of a former period. Millions of years were required for the transformation; but the same change is in progress now, where peat beds are forming from turf.

Coal is found in all countries, the largest beds being in the United States. From the nature of its formation, coal varies much in purity.

Anthracite, or hard coal, is purest in carbon, some varieties having from 90 to 95 per cent. This represents most complete distillation in the earth; *i.e.* the gases have mostly been driven off. It is much used in New England.

48. Bituminous, or soft coal, crocks the hands, and burns rapidly with much flame and smoke. The greater part of the coal in the earth is bituminous. It represents incomplete distillation. Hence, by artificially distilling it, illuminating gas is made. See page 180. It is far less pure C than anthracite.

49. Cannel Coal is a variety of bituminous coal which can be ignited like a candle. This is because so many of the gases are still left, and it shows cannel to be less pure C than bituminous coal.

50. Lignite, Peat, Turf, etc., are still less pure varieties of C. Construct a table of the naturally occurring forms of this element, in the order of their purity.

Carbon forms the basis of all vegetable and animal life; it is found in many rocks, mineral oils, asphaltum, natural gas, and in the air as CO_2.

51. C a Reducing Agent.

Experiment 31.— Put into a small ignition-tube a mixture of 4 or 5^g of powdered copper oxide (CuO), with half its bulk of powdered charcoal. Heat strongly for ten or fifteen minutes. Examine the contents for metallic copper. With which element of CuO has C united? The reaction may be written: $CuO + C = CO + Cu$. Complete and explain.

A **Reducing, or Deoxidizing, Agent** is a substance which takes away oxygen from a compound. C is the most common and important reducing agent, being used for this purpose in smelting iron and other ores, making water-gas, etc.

An **Oxidizing Agent** is a substance that gives up its O to a reducing agent. What oxidizing agent in the above experiment?

52. C a Decolorizer.

Experiment 32.— Put 3 or 4^g of bone-black into a receiver, and add 10 or 15^{cc} of cochineal solution. Shake this thoroughly, covering the bottle with the hand. Then pour the whole on a filter paper, and examine the filtrate. If all the color is not removed, filter again. What property of C is shown by this experiment? Any other coloring solution may be tried.

The decolorizing power of charcoal is an important characteristic. Animal charcoal is used in large quantities for decolorizing sugar. The coloring matter is taken out mechanically by the C, there being no chemical action.

53. C a Disinfectant.

Experiment 33. — Repeat the previous experiment, adding a solution of H_2S, *i.e.* hydrogen sulphide, in water, instead of cochineal solution. See page 120. Note whether the bad odor is removed. If not, repeat.

Charcoal has the property of absorbing large quantities of many gases. Ill-smelling and noxious gases are condensed in the pores of the C; O is taken in at the same time from the air, and these gases are there oxidized and rendered odorless and harmless. For this reason charcoal is much used in hospitals and sick-rooms, as a disinfectant. This property of condensing O, as well as other gases, is shown in the experiment below.

54. C an Absorber of Gases and a Retainer of Heat.

Experiment 34. — Put a piece of phosphorus of the size of a pea, and well dried, on a thick paper. Cover it well with bone-black, and look for combustion after a while. O has been condensed from the air, absorbed by the C, and thus communicated to the P. Burn all the P at last.

CHAPTER XI.

VALENCE.

55. The Symbols NaCl and MgCl$_2$ Differ in Two Ways. — What are they? Let us see why the atom of Mg unites with two Cl atoms, while that of Na takes but one. If the atoms of two elements attract each other, there must be either a general attraction all over their surfaces, or else some one or more *points* of attraction. Suppose the latter to be true, each atom must have one or more poles or bonds of attraction, like the poles of a magnet. Different elements differ in their number of bonds. Na has one, which may be written graphically Na—; Cl has one, —Cl. When Na unites with Cl, the bonds of each element balance, as follows: Na—Cl. The element Mg, however, has two such bonds, as Mg= or —Mg—. When Mg unites with Cl, in order to balance, or saturate, the bonds, it is evident that two atoms of Cl must be used, as $Mg = {Cl, \atop Cl,}$ or Cl—Mg—Cl, or MgCl$_2$.

A compound or an element, in order to exist, must have no free bonds. In organic chemistry the exceptions to this rule are very numerous, and, in fact, we do not know that atoms have bonds at all; but we can best explain the phenomena by supposing them, and for a general statement we may say that there must be no free bonds. In binaries the bonds of each element must balance.

56. The Valence, Quantivalence, of an Element is its Combining Power Measured by Bonds. — H, having the least number of bonds, one, is taken as the unit. Valence has always to be taken into account in writing the symbol of a compound. It is often written above and after the elements, as K^I, Mg^{II}.

An element having a valence of one is a monad; of two, a dyad; three, a triad; four, tetrad; five, pentad; six, hexad, etc. It is also said to be monovalent, di- or bivalent, etc. This theory of bonds shows why an atom cannot exist alone. It would have free or unused bonds, and hence must combine with its fellow to form a molecule, in case of an element as well as in that of a compound. This is illustrated by these graphic symbols in which there are no free bonds: $H-H$, $O=O$, $N\equiv N$, $C\equiv C$. A graphic symbol shows apparent molecular structure.

After all, how do we know that there are twice as many Cl atoms in the chloride of magnesium as in that of sodium? The compounds have been analyzed over and over again, and have been found to correspond to the symbols $MgCl_2$ and $NaCl$. This will be better understood after studying the chapter on atomic weights. In writing the symbol for the union of H with O, if we take an atom of each, the bonds do not balance, $H-=O$, the former having one; the latter, two. Evidently two atoms of H are needed, as $H-O-H$, or $\genfrac{}{}{0pt}{}{H}{H}=O$, or H_2O. In the union of Zn and O, each has two bonds; hence they unite atom with atom, $Zn=O$, or ZnO.

Write the graphic and the common symbols for the union of H^I and Cl^I; of K^I and Br^I; Ag^I and O^{II}; Na^I and S^{II}; H^I and P^{III}. Study valences, page 12. It will be seen that some elements have a variable quantivalence. Sn has either 2 or 4; P has 3 or 5. It usually varies by two for a given element, as though a pair of bonds sometimes saturated each other; e.g. =Sn=, a quantivalence of 4, and |Sn=, a quantivalence of 2. There are, therefore, two oxides of tin, SnO and SnO_2, or Sn=O and O=Sn=O. Write symbols for the two chlorides of tin; two oxides of P (page 12); two oxides of arsenic.

The chlorides of iron are $FeCl_2$ and Fe_2Cl_6. In the latter, it might be supposed that the quantivalence of Fe is 3, but the graphic symbol shows it to be 4. It is called a pseudo-triad, or false triad. Cr and Al are also pseudo-triads.

$$\begin{array}{cc} Cl & Cl \\ | & | \\ Cl-Fe-Fe-Cl. \\ | & | \\ Cl & Cl \end{array}$$

Write formulæ for two oxides of iron; the oxide of Al.

57. A Radical is a Group of Elements which has no separate existence, but enters into combination like a single atom; e.g. (NO_3) in the compounds HNO_3 or KNO_3; (SO_4) in H_2SO_4. In HNO_3 the radical has a valence of 1, to balance that of H, $H-(NO_3)$. In H_2SO_4, what is the valence of (SO_4)? Give it in each of these radicals, noting first that of the first element: $K(NO_3)$, $Na_2(SO_4)$, $Na_2(CO_3)$, $K(ClO_3)$, $H_3(PO_4)$, $Ca_3(PO_4)_2$, $Na_4(SiO_4)$.

Suppose we wish to know the symbol for calcium phosphate. Ca and PO_4 are the two parts. In $H_3(PO_4)$ the radical is a triad, to balance H_3. Ca is a dyad, $Ca=\equiv(PO_4)$. The least common multiple of the bonds (2 and 3) is 6, which, divided by 2 (no. Ca bonds), gives 3 (no. Ca atoms to be taken). $6 \div 3$ (no. (PO_4) bonds) gives 2 (no. PO_4 radicals to be taken). Hence the symbol $Ca_3(PO_4)_2$. Verify this by writing graphically.

Write symbols for the union of Mg and (SO_4), Na and (PO_4), Zn and (NO_3), K and (NO_3), K and (SO_4), Mg and (PO_4), Fe and (SO_4) (both valences of Fe), Fe and (NO_3), taking the valences of the radicals from HNO_3, H_2SO_4, H_3PO_4.

CHAPTER XII.

ELECTRO-CHEMICAL RELATION OF ELEMENTS.

58. Examine untarnished pieces of iron, silver, nickel, lead, etc.; also quartz, resin, silk, wood, paper. Notice that from the first four light is reflected in a different way from that of the others. This property of reflecting light is known as luster. Metals have a metallic luster which is peculiar to themselves; and this, for the present, may be regarded as their chief characteristic. Are they at the positive or negative end of the list? See page 43. How is it with the non-metals? This arrangement has a significance in chemistry which we must now examine. The three appended experiments show how one metal can be withdrawn from solution by a second, this second by a third, the third by a fourth, and so on. For expedition, three pupils can work together for the three following experiments, each doing one, and examining the results of the others.

59. Deposition of Silver.

Experiment 35. — Put a ten-cent Ag coin into an evaporating-dish, and pour over it a mixture of 5^{cc} HNO_3 and 10^{cc} H_2O. Warm till all, or nearly all, the Ag dissolves. Remove the lamp.

$3 Ag + 4 HNO_3 = 3 AgNO_3 + 2 H_2O + NO$. Then add 10^{cc} H_2O, and at once put in a short piece of Cu wire, or a cent. Leave till quite a deposit appears, then pour off the liquid, wash the deposit thoroughly, and remove it from the coin. See whether the metal resembles Ag. $2 AgNO_3 + Cu = ?$

60. Deposition of Copper.

Experiment 36. — Dissolve a cent or some Cu turnings in dilute HNO_3, as in Experiment 35, and dilute the solution. $3 Cu + 8 HNO_3 = 3 Cu(NO_3)_2 + 4 H_2O + 2 NO$.

Then put in a clean strip of Pb, and set aside as before, examining the deposit finally. $Cu(NO_3)_2 + Pb = ?$

61. Deposition of Lead.

Experiment 37. — Perform this experiment in the same manner as the two previous ones, dissolving a small piece of Pb, and using a strip of Zn to precipitate the Pb. $3 Pb + 8 HNO_3 = 3 Pb(NO_3)_2 + 4 H_2O + 2 NO$. $Pb(NO_3)_2 + Zn = ?$

62. Explanation.

— These experiments show that Cu will replace Ag in a solution of $AgNO_3$, that Pb will replace and deposit Cu from a similar compound, and that Zn will deposit Pb in the same way. They show that the affinity of Zn for (NO_3) is stronger than either Ag, Cu, or Pb. We express this affinity by saying that Zn is the most positive of the four metals, while Ag is the most negative. Cu is positive to Ag, but negative to Pb and Zn. Which of the four elements are positive to Pb, and which negative? Mg would withdraw Zn from a similar solution, and be in its turn withdrawn by Na. The table on page 43 is founded on this relation. A given element is positive to every element above it in the list, and negative to all below it.

Metals are usually classed as positive, non-metals as negative. Each in union with O and H gives rise to a very important class of compounds, — the negative to acids, the positive to bases.

In the following, note whether the positive or the negative element is written first: HCl, Na_2O, As_2S_3, $MgBr_2$, Ag_2S. Na_2SO_4 is made up of two parts, Na_2 being posi-

ORDER.
—

Negative or Non-metallic Elements. Acid-forming with H (usually OH).

- Oxygen
- Sulphur
- Nitrogen
- Fluorine
- Chlorine
- Bromine
- Iodine
- Phosphorus
- Arsenic
- Carbon
- Silicon

Hydrogen

Positive or Metallic Elements. Base-forming with OH.

- Gold
- Platinum
- Mercury
- Silver
- Copper
- Tin
- Lead
- Iron
- Zinc
- Aluminium
- Magnesium
- Calcium
- Sodium
- Potassium

+

tive, the radical SO_4 negative. Like elements, radicals are either positive or negative. In the following, separate the positive element from the negative radical by a vertical line: Na_2CO_3, $NaNO_3$, $ZnSO_4$, $KClO_3$.

The most common positive radical is NH_4, ammonium, as in NH_4Cl. It always deports itself as a metal. The commonest radical is the negative OH, called hydroxyl, from hydrogen-oxygen. Take away H from the symbol of water, H—O—H, and hydroxyl —(OH) with one free bond is left. If an element takes the place of H, *i.e.* unites with OH, the compound is called a hydrate. KOH is potassium hydrate. Name NaOH, $Ca(OH)_2$, NH_4OH, $Zn(OH)_2$, $Al_2(OH)_6$. Is the first part of each symbol above positive or negative?

H has an intermediate place in the list. It is a constituent of both acids and bases, and of the neutral substance, water.

CHAPTER XIII.

ELECTROLYSIS.

The following experiment is to be performed only by the teacher, but pupils should make drawings and explain.

63. Decomposition of Water.

Experiment 38. — Arrange "in series" two or more cells of a Bunsen battery (Physics, page 164),[1] and attach the terminal wires to an electrolytic apparatus (Fig. 19) filled with water made slightly acid with H_2SO_4. Construct a diagram of the apparatus, marking the Zn in the liquid $+$, since it is positive, and the C, or other element, $-$. Mark the electrode attached to the Zn $-$, and that attached to the C $+$; positive electricity at one end of a body commonly implies negative at the other. Opposites attract, while like electricities repel each other. These analogies will aid the memory. At the $+$ electrode is the $-$ element of H_2O, and at the $-$ electrode the $+$ element. Note, page 43, whether H or O is positive with reference to the other, and write the symbol for each at the proper electrode. Compare the diagram with the apparatus, to verify your conclusion. Why does gas collect twice as fast at one electrode as at the other? What does this prove of the composition of water? When filled, test the gases in each tube, for O and H, with a burning stick. Electrical analysis is called *electrolysis*.

Fig. 19.

If a solution of NaCl be electrolyzed, which element will go to the $+$ pole? Which, if the salt were K_2SO_4?

[1] References are made in this book to Gage's Introduction to Physical Science.

Explain these reactions in the electrolysis of that salt. $K_2SO_4 = K_2 + SO_3 + O$. SO_4 is unstable, and breaks up into SO_3 and O. Both K and SO_3 have great affinity for water. $K_2 + 2 H_2O = 2 KOH + H_2$. $SO_3 + H_2O = H_2SO_4$.

The base KOH would be found at the — electrode, and the acid H_2SO_4 at the + electrode.

The positive portion, K, uniting with H_2O forms a base; the negative part, SO_3, with H_2O forms an acid. Of what does this show a salt to be composed?

64. Conclusions. — These experiments show (1) that at the + electrode there always appears the negative element, or radical, of the compound, and at the — electrode the positive element; (2) that these elements unite with those of water, to make, in the former case, acids, in the latter, bases; (3) that acids and bases differ as negative and positive elements differ, each being united with O and H, and yet producing compounds of a directly opposite character; (4) that salts are really compounded of acids and bases. This explains why salts are usually inactive and neutral in character, while acids and bases are active agents. Thus we see why the most positive or the most negative elements in general have the strongest affinities, while those intermediate in the list are inactive, and have weak affinities; why alloys of the metals are weak compounds; why a neutral substance, like water, has such a weak affinity for the salts which it holds in solution; and why an aqueous solution is regarded as a mechanical mixture rather than a chemical compound. In this view, the division line between chemistry and physics is not a distinct one. These will be better understood after studying the chapters on acids, bases and salts.

CHAPTER XIV.

UNION BY VOLUME.

65. Avogadro's Law of Gases. — Equal volumes of all gases, the temperature and pressure being the same, have the same number of molecules. This law is the foundation of modern chemistry. A cubic centimeter of O has as many molecules as a cubic centimeter of H, a liter of N the same number as a liter of steam, under similar conditions. Compare the number of molecules in 5^l of N_2O with that in 10^l Cl. 7^{cc} vapor of I to 6^{cc} vapor of S. The half-molecules of two gases have, of course, the same relation to each other, and in elements the half-molecule is usually the atom.

The molecular volumes — molecules and the surrounding space — of all gases must therefore be equal, as must the half-volumes. Notice that this law applies only to gases, not to liquids or solids. Let us apply it to the experiment for the electrolysis of water. In this we found twice as much H by volume as O. Evidently, then, steam has twice as many molecules of H as of O, and twice as many half-molecules, or atoms. If the molecule has one atom of O, it must have two of H, and the formula will be H_2O.

Suppose we reverse the process and synthesize steam, which can be done by passing an electric spark through a mixture of H and O in a eudiometer over mercury; we should need to take twice as much H as O. Now when 2^{cc} of H combine thus with 1^{cc} of O, only 2^{cc} of steam are produced. Three volumes are condensed into two volumes, and of course three molecular volumes into two, three atomic volumes into two. This may be written as follows: —

$$\square + \square + \square = \square$$
$$H + H + O = H_2O.$$

This is a condensation of one-third.

If 2^l of chlorhydric acid gas be analyzed, there will result 1^l of H and 1^l of Cl. The same relation exists between the molecules and the atoms, and the reaction is: —

$$\square = \square + \square$$
$$HCl = H + Cl.$$

Reverse the process, and 1^l of H unites with 1^l of Cl to produce 2^l of the acid gas; there is no condensation, and the symbol is HCl. In seven volumes HCl how many of each constituent?

The combination of two volumes of H with one volume of S is found to produce two volumes of hydrogen sulphide. Therefore two atoms of H combine with one of S to form a molecule whose symbol is H_2S.

$$\square + \square + \square = \square$$
$$H + H + S = H_2S.$$

What is the condensation in this case?

PROBLEMS.

(1) How many liters of S will it take to unite with 4^l of H? How much H_2S will be formed?

(2) How many liters of H will it take to combine with 5^l of S? How much H_2S results?

(3) In 6^l H_2S how many liters H, and how much S? Prove.

(4) In four volumes H_2S how many volumes of each constituent?

(5) If three volumes of H be mixed with two volumes of S, so as to make H_2S, how much will be formed? How much of either element will be left?

An analysis of 2^{cc} of ammonia gives 1^{cc} N and 3^{cc} H. The symbol must then be NH_3, the reaction, —

$$\square = \square + \square + \square + \square$$
$$NH_3 = N + H + H + H.$$

What condensation in the synthesis of NH_3?

In 12^{cc} NH_3 how many cubic centimeters of each element? In $2\frac{1}{2}^{cc}$? How much H by volume is required to combine with nine volumes of N? How many volumes of NH_3 are produced?

In elements that have not been weighed in the gaseous state, as C, the evidence of atomic volume is not direct, but we will assume it. Thus two volumes of marsh gas would separate into one of C and four of H. What is its symbol and supposed condensation? Two volumes of alcohol vapor resolve into two of C, six of H, and one of O. What is its symbol? its condensation?

The symbol itself of a compound will usually show what its condensation is; *e.g.* HCl, HBr, HF, etc., have two atoms; hence there will be no shrinkage. In H_2O, SO_2, CO_2, the molecule has three atoms condensed into the space of two, or one-third shrinkage. In NH_3 four volumes are crowded into the space of two, a condensation of one-half.

P, As, Hg, Zn, have exceptional atomic volumes. See page 112.

CHAPTER XV.

ACIDS AND BASES.

66. What Acids Are.

Experiment 39.— Pour a few drops of chlorhydric acid, HCl, into a clean evaporating-dish. Add 5^{cc} H$_2$O, and stir. Touch a drop to the tongue, noting the taste. Dip into it the end of a piece of blue litmus paper, and record the result. Thoroughly wash the dish, then pour in a few drops of nitric acid, HNO$_3$, and 5^{cc} H$_2$O, and stir. Taste, and test with blue litmus. Test in the same way sulphuric acid, H$_2$SO$_4$. Name two characteristics of an acid. In a vertical line write the formulæ of the acids above. What element is common to them all? Is the rest of the formula positive or negative?

67. An Acid is a substance composed of H and a negative element or radical. It has usually a sour taste, and turns blue litmus red. Litmus is a vegetable extract obtained from lichens in Southern Europe. Acids have the same action on many other vegetable pigments. Are the following acid formulæ, and why? H$_2$SO$_3$, HBr, HNO$_2$, H$_3$PO$_3$, H$_4$SiO$_4$. Most acids have O as well as H. Complete the symbols for acids in the following list, and name them, from the type given:—

HCl, chlorhydric acid.   HNO$_3$, nitric acid.
? Br, ?   ? Cl ? ?
? I, ?   ? Br ? ?
? F, ?   ? I ? ?
H$_3$PO$_4$, phosphoric acid.   H$_3$PO$_3$, phosphorous acid.
? As ? ?   ? As ? ?

ACIDS AND BASES.

Complete these equations: —

$H_2SO_3 - H_2O = ?$

$H_2SO_4 - H_2O = ?$

$H_2CO_3 - H_2O = ?$

$2 HNO_3 - H_2O = ?$

$2 HNO_2 - H_2O = ?$

$2 H_3AsO_4 - 3 H_2O = ?$

Are the products in each case metallic or non-metallic oxides? They are called anhydrides. Notice that each is formed by the withdrawal of water from an acid. Reverse the equations; as, $SO_3 + H_2O = ?$

68. An Anhydride is what remains after water has been removed from an acid; or, it is the oxide of a non-metallic element, which, united with water, forms an acid. SO_2 is sulphurous anhydride, SO_3 sulphuric anhydride, the ending *ic* meaning more O, or negative element, than *ous*. Name the others above.

Anhydrides were formerly called acids, — anhydrous acids, in distinction from hydrated ones, as CO_2 even now is often called carbonic acid.

Experiment 40. — Hold a piece of wet blue litmus paper in the fumes of SO_2, and note the acid test. Try the same with dry litmus paper.

Experiment 41. — Burn a little S in a receiver of air containing 10^{cc} H_2O, and loosely covered, as in the O experiment. Then shake to dissolve the SO_2. $H_2O + SO_2 = H_2SO_3$. Apply test paper.

69. Naming Acids. — Compare formulæ H_2SO_3 and H_2SO_4. Of two acids having the same elements, the name of the one with least O, or negative element, ends in *ous*, the other in *ic*. H_2SO_3 is sulphurous acid; H_2SO_4, sulphuric acid. Name H_3PO_4 and H_3PO_3; H_3AsO_3 and H_3AsO_4; HNO_2 and HNO_3.

If there are more than two acids in a series, the prefixes *hypo*, less, and *per*, more, are used. The following is such a series: $HClO$, $HClO_2$, $HClO_3$, $HClO_4$.

$HClO_3$ is chlor*ic* acid; $HClO_2$, chlor*ous*; $HClO$, *hypo-*

chlor*ous*; HClO₄ *per*chlor*ic*. *Hypo* means less of the negative element than *ous*; *per* means more of the negative element than *ic*. Name: H₃PO₄ (*ic*), H₃PO₃, H₃PO₂. Also HBrO (HBrO₂ does not exist), HBrO₃ (*ic*), HBrO₄.

What are the three most negative elements? Note their occurrence in the three strongest and most common acids. Hereafter note the names and symbols of all the acids you see.

70. What Bases Are.

Experiment 42. — Put a few drops of NH₄OH into an evaporating-dish. Add 5^{cc} H₂O, and stir. Taste a drop. Dip into it a piece of red litmus paper, noting the effect. Cleanse the dish, and treat in the same way a few drops NaOH solution, recording the result. Do the same with KOH. Acid stains on the clothing, with the exception of those made by HNO₃, may be removed by NH₄OH. H₂SO₄, however, rapidly destroys the fiber of the cloth.

Name two characteristics of a base. In the formulæ of those bases, what two common elements? Name the radical. Compare those symbols with the symbol for water, HOH. Is (OH) positive or negative? Is the other part of each formula positive or negative? What are two constituents, then, of a base? Bases are called hydrates. Write in a vertical line five positive elements. Note the valence of each, and complete the formula for its base. Affix the names. Can you see any reason why the three bases above given are the strongest? See page 43.

Taking the valences of Cr and Fe, write symbols for two sets of hydrates, and name them. Try to recognize and name every base hereafter met with.

A **Base** is a substance which is composed of a metal, or positive radical, and OH. It generally turns red litmus blue, and often has an acrid taste.

An Alkali is a base which is readily soluble in water. The three principal alkalies are NH_4OH, KOH, and $NaOH$.

Alkali Metals are those which form alkalies. Name three.

An Alkaline Reaction is the turning of red litmus blue.

An Acid Reaction is the turning of blue litmus red.

Experiment 43. — Pour 5^{cc} of a solution of litmus in water, into a clean t.t. or small beaker. Pour 2 or 3^{cc} of HCl into an evaporating-dish, and the same quantity of NH_4OH into another dish. Take a drop of the HCl on a stirring-rod and stir the litmus solution with it. Note the acid reaction. Clean the rod, and with it take a drop (or more if necessary) of NH_4OH, and add this to the red litmus solution, noting the alkaline reaction. . Experiment in the same way with the two other principal acids and the two other alkalies.

Litmus paper is commonly used to test these reactions, and hereafter whenever the term *litmus* is employed in that sense, the test-paper should be understood. This paper can be prepared by dipping unglazed paper into a strong aqueous solution of litmus.

CHAPTER XVI.

SALTS.

71. Acids and Bases are usually Opposite in Character. — When two forces act in opposition they tend to neutralize each other. We may see an analogy to this in the union of the two opposite classes of compounds, acids and bases, to form salts.

72. Neutralization.

Experiment 44. — Put into an evaporating-dish 5^{cc} of NaOH solution (page 208). Add HCl to this from a t.t., a few drops at a time, stirring the mixture with a glass rod (Fig. 20), and testing it with litmus paper, until the liquid is neutral, *i.e.* will not turn the test paper from blue to red, or red to blue. Test with both colors. If it turns blue to red, too much acid has been added; if red to blue, too much base. When it is very nearly neutral, add the reagent, HCl

Fig. 20.

Fig. 21.

or NaOH, a drop at a time with the stirring-rod. It must be absolutely neutral to both colors. Evaporate the water by heating the dish over asbestus paper, wire gauze, or sand, in an iron plate (Fig. 21) till the residue becomes dry and white. Cool the residue, taste, and name it. The equation is: $HCl + NaOH = NaCl + HOH$ or H_2O. Note which elements, positive or negative, change places. Why was the liquid boiled? The residue is a type of a large class of compounds, called salts.

Experiment 45. — Experiment in the same way with KOH solution and H_2SO_4, applying the same tests. $H_2SO_4 + 2 KOH = K_2SO_4 + 2 HOH$. What is the solid product?

Experiment 46. — Neutralize NH_4OH with HNO_3, evaporate, apply the tests, and write the equation. Write equations for the combination of $NaOH$ and H_2SO_4; $NaOH$ and HNO_3; KOH and HCl; KOH and HNO_3; NH_4OH and HCl; NH_4OH and H_2SO_4. Describe the experiment represented by each equation, and be sure you can perform it if asked to do so. What is the usual action of a salt on litmus? How is a salt made? What else is formed at the same time? Have all salts a saline taste? Does every salt contain a positive element or radical? A negative?

73. A Salt is the product of the union of a positive and a negative element or radical; it may be made by mixing a base and an acid.

The salt KI represents what acid? What base, or hydrate? Write the equation for making KI from its acid and base. Describe the experiment in full. Classify, as to acids, bases, or salts: KBr, $Fe(OH)_2$, HI, $NaBr$, HNO_2, $Al_2(OH)_6$, $KClO_3$, $HClO_3$, H_2S, K_2S, H_2SO_3, K_2SO_4, $Ca(OH)_2$, $CaCO_3$, $NaBrO_3$, $CaSO_4$, H_2CO_3, K_2CO_3, $Cu(OH)_2$, $Cu(NO_3)_2$, $PbSO_4$, H_3PO_4, Na_3PO_4. In the *salts* above, draw a light vertical line, separating the positive from the negative part of the symbol. Now state what acid each represents. What base. Write the reaction in the preparation of each salt above from its acid and base; then state the experiment for producing it.

74. Naming Salts. — (NO_3) is the nitrate radical; KNO_3 is potassium nitrate. From what acid? (NO_2) is the nitrite radical; KNO_2 is potassium nitrite. From what acid? Note that the endings of the acids are *ous* and *ic;* also that the names of their salts end in *ite* and *ate*. From which acid — *ic* or *ous* — is the salt ending in *ate* derived? That ending in *ite?*

Name these salts, the acids from which they are derived, and the endings of both acids and salts: $NaNO_3$, $NaNO_2$, K_2SO_4, K_2SO_3, $CaSO_4$, $CaSO_3$, $KClO_3$, $KClO_2$, $KClO$, $KClO_4$ (use prefixes *hypo* and *per*, as with acids), $Ca_3(PO_4)_2$, $Ca_3(PO_3)_2$, $CuSO_4$, $CuSO_3$, $AgNO_3$, $Cu(NO_3)_2$. FeS, FeS_2, are respectively *ferrous sulphide* and *ferric sulphide*. Name: $HgCl$, $HgCl_2$, $FeCl_2$, Fe_2Cl_6, $FeSO_4$, $Fe_2(SO_4)_3$.

75. Acid Salts. — Write symbols for nitric, sulphuric, phosphoric acids. How many H atoms in each? Replace all the H in the symbol of each with Na, and name the products. Again, in sulphuric acid replace one atom of H with Na; then in phosphoric replace first one, then two, and finally three H atoms with Na. $HNaSO_4$ is hydrogen sodium sulphate; HNa_2PO_4 is hydrogen di-sodium phosphate. Name the other salts symbolized. Name $HNaNH_4PO_4$. Though these products are all salts, some contain replaceable H, and are called acid salts. Those which have all the H replaced by a metal are normal salts. Name and classify, as to normal or acid salts: Na_2CO_3, $HNaCO_3$, K_2SO_4, $HKSO_4$, $(NH_4)_2SO_4$, HNH_4SO_4, Na_3PO_4, HNa_2PO_4, H_2NaPO_4.

The *basicity* of an acid is determined by the number of replaceable H atoms in its molecule. It is called *monobasic* if it has one; *dibasic* if two; *tri-* if three, etc. Note the basicity of each acid named above. How many possible salts of H_2SO_4 with Na? Of H_3PO_4 with Na? Which are normal and which acid? What is the basicity of H_4SiO_4?

Some normal, as well as acid, salts change litmus. Na_2CO_3, representing a strong base and a weak acid, turns it blue. There are other modes of obtaining salts, but this is the only one which we shall consider.

76. Salts Occur Abundantly in Nature, such as NaCl, $MgSO_4$, $CaCO_3$. Acids and bases are found in small quantities only. Why is this? Why are there not springs of H_2SO_4 and NH_4OH? We have seen that acids and bases are extremely active, have opposite characters, and combine to form relatively inactive salts. If they existed in the free state, they would soon combine by reason of their strong affinities. This is what in all ages of the world has taken place, and this is why salts are common, acids and bases rare. Active agents rarely exist in the free state in large quantities. Oxygen seems to be an exception, but this is because there is a superabundance of it. While vast quantities are locked up in compounds in rocks, water, and salts of the earth, much remains with which there is nothing to combine.

CHAPTER XVII.

CHLORHYDRIC ACID.

77. We have seen that salts are made by the union of acids and bases. Can these last be obtained from salts?

78. Preparation of HCl.

Experiment 47. — Into a flask put 10^g coarse NaCl, and add 20^{cc} H_2SO_4. Connect with Woulff bottles[1] partly filled with water, as in Figure 22. One bottle is enough to collect the HCl; but in that case it is less pure, since some H_2SO_4 and other impurities are carried over. Several may be connected, as in Figure 23. The water in the first bottle must be nearly saturated before much gas will pass into the second. Heat the mixture 15 or 20 minutes, not very strongly, to prevent too much foaming. Notice any current in the first bottle. $NaCl + H_2SO_4 = HNaSO_4 + HCl$. Intense heat would have given: $2\,NaCl + H_2SO_4 = Na_2SO_4 + 2\,HCl$. Compare these equations with those for HNO_3. In which equation above is H_2SO_4 used most economically? Both reactions take place when HCl is made on the large scale.

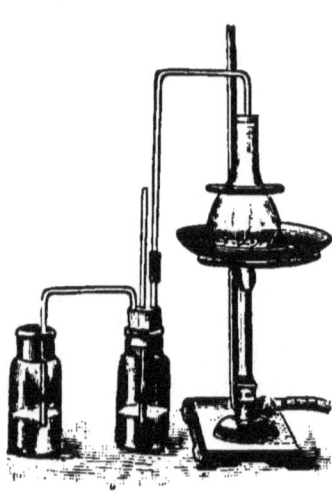

Fig. 22.

79. Tests.

Experiment 48. — (1) Test with litmus the liquid in each Woulff

---

[1] Woulff bottles may be made by fitting to wide-mouthed bottles corks with three holes, through which pass two delivery tubes, and a central safety tube dipping into the liquid, as in Figures 22 and 23.

bottle. (2) Put a piece of Zn into a t.t. and cover it with liquid from the first bottle. Write the reaction, and test the gas. (3) To 2^{cc} solution AgNO$_3$ (page 208) in a t.t. add 2^{cc} of the acid. Describe, and write the reaction. Is AgCl soluble in water? (4) Into a t.t. pour 5^{cc} Pb(NO$_3$)$_2$ solution, and add the same amount of prepared acid. Give the description and the reaction. (5) In the same way test the acid with Hg$_2$(NO$_3$)$_2$ solution, giving the reaction. (6) Make a little HCl in a t.t., and bring the gas escaping from the d.t. in contact with a burning stick. Does it support the combustion of C? (7) Hold a piece of dry litmus paper against it.

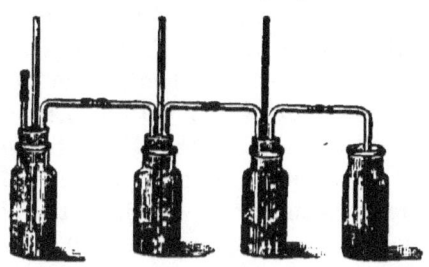

Fig. 23.

(8) Hold it over 2^{cc} of NH$_4$OH in an evaporating-dish. Describe, name the product, and write the reaction. (3), (4), (5), (8), are characteristic tests for this acid.

80. Chlorhydric, Hydrochloric or Muriatic, Acid is a Gas. — As used, it is dissolved in water, for which it has great affinity. Water will hold, according to temperature, from 400 to 500 times its volume of HCl. Hundreds of thousands of tons of the acid are annually made, mostly in Europe, as a bye-product in Na$_2$CO$_3$ manufacture. The gas is passed into towers through which a spray of water falls; this absorbs it. The yellow color in most commercial HCl indicates impurities, some of which are Fe, S, As, and organic matter. As, S, etc., come from the pyrites used in making H$_2$SO$_4$. Chemically pure (C.P.) acid is freed from these, and is without color. The gas may be dried by passing it through a glass tube holding CaCl$_2$ (Fig. 16) and collecting it over mercury.

The muriatic acid of commerce consists of about two-thirds water by weight. HCl can also be made by direct union of its constituents (page 100).

81. Uses. — HCl is used to make Cl, and also bleaching-powder. Its use as a reagent in the laboratory is illustrated by the following experiment: —

Experiment 49. — Put into a t.t. 2^{cc} $AgNO_3$ solution, add 5^{cc} H_2O, then add slowly HCl so long as a ppt. (precipitate) is formed. This ppt. is AgCl. Now in another t.t. put 2^{cc} $Cu(NO_3)_2$ solution, add 5^{cc} H_2O, then a little HCl. No ppt. is formed. Now if a solution of $AgNO_3$ and a solution of $Cu(NO_3)_2$ were mixed, and HCl added, it is evident that the silver would be precipitated as chloride of silver, while the copper would remain in solution. If now this be filtered, the silver will remain on the filter paper, while in the filtrate will be the copper. Thus we shall have performed an analysis, or separated one metal from another. Perform it. Note, however, that any soluble chloride, as NaCl, would produce the same result as HCl.

BROMHYDRIC AND IODIHYDRIC ACIDS.

82. NaCl, being the most abundant compound of Cl, is the source of commercial HCl. KCl treated in the same way would give a like product. Theoretically HBr and HI might be made in the same way from NaBr and NaI, but the affinity of H for Br and I is weak, and the acids separate into their elements, when thus prepared.

83. To make HI.

Experiment 50. — Drop into a t.t. three or four crystals of I, and add 10^{cc} H_2O. Hold in the water the end of a d.t. from which H_2S gas is escaping (page 120). Observe any deposit, and write the reaction.

FLUORHYDRIC ACID.

84. Preparation and Action.

Experiment 51. — Put 3 or 4^g powdered CaF_2, *i.e.* fluor spar or fluorite, into a shallow lead tray, *e.g.* 4×5^{cm}, and pour over it 4 or 5^{cc} H_2SO_4. A piece of glass large enough to cover this should previously be warmed and covered on one side with a very thin coat of beeswax. To distribute it evenly, warm the other side of the glass over a flame. When cool, scratch a design (Fig. 24) through the wax with a sharp metallic point. Lay the glass, film side down, over the lead tray. Warm this five minutes or more

Fig. 24.

by placing it high over a small flame (Fig. 25) to avoid melting the wax. Do not inhale the fumes. Take away the lamp, and leave the tray and glass where it is not cold, for half an hour or more. Then remove the wax and clean the glass with naphtha or benzine. Look for the etching.

Fig. 25.

Two things should have occurred: (1) the generation of HF. Write the equation for it. (2) Its etching action on glass. In this last process HF acts on SiO_2 of the glass, forming H_2O and SiF_4. Why cannot HF be kept in glass bottles?

A dilute solution of HF, which is a gas, may be kept in gutta percha bottles, the anhydrous acid in platinum only; but for the most part, it is used as soon as made, its chief use being to etch designs on glass-ware. Glass is also often etched by a blast of sand (SiO_2).

Notice the absence of O in the acids HF, HCl, HBr, HI, and that each is a gas. HF is the only acid that will dissolve or act appreciably on glass.

CHAPTER XVIII.

NITRIC ACID.

85. Preparation.

Experiment 52.—To 10^g KNO_3 or $NaNO_3$, in a flask, add 15^{cc} H_2SO_4. Securely fasten the cork of the d.t., as HNO_3 is likely to loosen it, and pass the other end to the bottom of a t.t. held deep in a bottle of water (Fig. 26). Apply heat, and collect 4 or 5^{cc} of the liquid. The

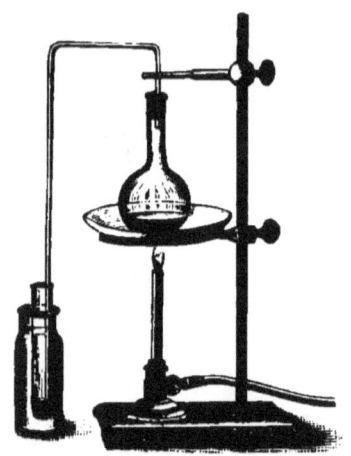
Fig. 26.

Fig. 27.

usual reaction is: $KNO_3 + H_2SO_4 = HKSO_4 + HNO_3$. With greater heat, $2 KNO_3 + H_2SO_4 = K_2SO_4 + 2 HNO_3$. Which is most economical of KNO_3? Of H_2SO_4? Instead of a flask, a t.t. may be used if desired (Fig. 27).

86. Properties and Tests.

Experiment 53.—(1) Note the color of the prepared liquid. (2) Put a drop on the finger; then wash it off at once. (3) Dip a quill or piece of white silk into it; then wash off the acid. What color is imparted to animal substances? (4) Add a little to a few bits of Cu turnings, or to a Cu coin. Write the equation (page 42). (5) To 2^{cc} indigo solution add 2^{cc} HNO_3. State the leading properties of HNO_3 from these tests.

87. Chemically Pure HNO_3 is a Colorless Liquid. — The yellow color of that prepared in Experiment 52 is due to liquid NO_2 dissolved in it. It is then called fuming HNO_3, and is very strong. NO_2 is formed at a high temperature.

Commercial or ordinary HNO_3 is made from $NaNO_3$, this being cheaper than KNO_3; it is about half water.

88. Uses. — HNO_3 is the basis of many nitrates, as $AgNO_3$, used for photography, $Ba(NO_3)_2$ and $Sr(NO_3)_2$ for fire-works, and others for dyeing and printing calico; it is employed in making aqua regia, sulphuric acid, nitro-glycerine, gun-cotton, aniline colors, zylonite, etc.

Enough experiments have been performed to answer the question whether some acids can be prepared from their salts. H_2SO_4 is not so made, because no acid is strong enough to act on its salts. In making HCl, HNO_3, etc., sulphuric acid was used, being the strongest.

AQUA REGIA.

89. Preparation and Action.

Experiment 54. — Into a t.t. put 2^{cc} HNO_3 and 1^{qcm} of either Au leaf or Pt. Warm in a flame. If the metal is pure, no action takes place. Into another tube put 6^{cc} HCl and add a similar leaf. Heat this also. There should be no action. Pour the contents of one t.t. into the other. Note the effect. Which is stronger, one of the acids, or the combination of the two? Note the odor. It is that of Cl. $3 HCl + HNO_3 = NOCl + 2 H_2O + Cl_2$. This reaction is approximate only. The strength is owing to nascent chlorine, which unites with Au. $Au + 3 Cl = AuCl_3$. If Pt be used, $PtCl_4$ is produced.

No other acid except nitro-hydrochloric will dissolve Au or Pt; hence the ancients called it *aqua regia*, or king of liquids. It must be made as wanted, since it cannot be kept and retain its strength.

CHAPTER XIX.

SULPHURIC ACID.

90. Preparation.

Experiment 55. — Having fitted a cork with four or five perforations to a large t.t., pass a d.t. from three of these to three smaller t.t., leaving the others open to the air, as in Figure 28. Into one t.t. put 5^{cc} H$_2$O, into another 5^g Cu turnings and 10^{cc} H$_2$SO$_4$, into the third 5^g Cu turnings and 10^{cc} dilute HNO$_3$, half water. Hang on a ring stand, and slowly heat the tubes containing H$_2$O and H$_2$SO$_4$. Notice the fumes that pass into the large t.t.

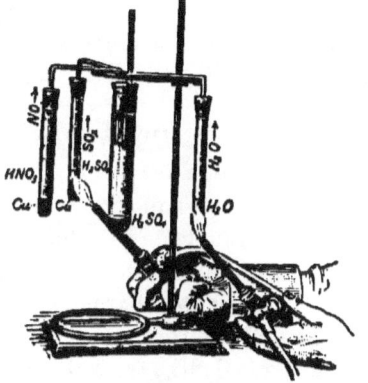

Fig. 28.

Trace out and apply to Figure 28 these reactions: —

(1) $Cu + 2 H_2SO_4 = CuSO_4 + 2 H_2O + SO_2$.

(2) $3 Cu + 8 HNO_3 = 3 Cu (NO_3)_2 + 4 H_2O + 2 NO$.

(3) $NO + O = NO_2$.

(4) $SO_2 + H_2O + NO_2 = H_2SO_4 + NO$.

(4) comes from combining the gaseous products in (1), (2), (3). In (3), NO takes an atom of O from the air, becoming NO$_2$, and at once gives it up to the H$_2$SO$_3$ (H$_2$O + SO$_2$), making H$_2$SO$_4$, and again goes through the same operation of taking up O and passing it along. NO is thus called a carrier of O. It is a reducing agent, while NO$_2$ is an oxidizing agent. This is a continuous process,

and very important, since it changes useless H_2SO_3 into valuable H_2SO_4. If exposed to the air, H_2SO_3 would very slowly take up O and become H_2SO_4.

Instead of the last experiment, this may be employed if preferred: Burn a little S in a receiver. Put into an evaporating-dish, 5^{cc} HNO_3, and dip a paper or piece of cloth into it. Hang the paper in the receiver of SO_2, letting no HNO_3 drop from it. Continue this operation till a small quantity of liquid is found in the bottle. The fumes show that HNO_3 has lost O. $2 HNO_3 + SO_2 = H_2SO_4 + 2 NO_2$.

91. Tests for H_2SO_4.

Experiment 56. — (1) Test the liquid with litmus. (2) Transfer it to a t.t., and add an equal volume of $BaCl_2$ solution. $H_2SO_4 + BaCl_2 = ?$ Is $BaSO_4$ soluble? (3) Put one drop H_2SO_4 from the reagent bottle in 10^{cc} H_2O in a clean t.t., and add 1^{cc} $BaCl_2$ solution. Look for any cloudiness. This is the characteristic test for H_2SO_4 and soluble sulphates, and so delicate that one drop in a liter of H_2O can be detected. (4) Instead of H_2SO_4, try a little Na_2SO_4 solution. (5) Put two or three drops of strong H_2SO_4 on writing-paper, and evaporate, high over a flame, so as not to burn the paper. Examine it when dry. (6) Put a stick into a t.t. containing 2^{cc} H_2SO_4, and note the effect. (7) Review Experiment 5. (8) Into an e.d. pour 5^{cc} H_2O, and then 15^{cc} H_2SO_4. Stir it meantime with a small t.t. containing 2 or 3^{cc} NH_4OH, and notice what takes place in the latter; also note the heat of the e.d.

The effects of (5), (6), (7), and (8) are due to the intense affinity which H_2SO_4 has for H_2O. So thirsty is it that it even abstracts H and O from oxalic acid in the right proportion to form H_2O, combines them, and then absorbs the water.

92. Affinity for Water.
— This acid is a desiccator or dryer, and is used to take moisture from the air and prevent metallic substances from rusting. In this way it dilutes itself, and may increase its weight threefold. In diluting, the acid must always be poured into the water slowly and with stirring, not water into the acid, since, as H_2O is lighter than H_2SO_4, heat enough may be set free at the surface of contact to cause an explosion.

SULPHURIC ACID.   65

Contraction also takes place, as may be shown by accurately measuring each liquid in a graduate, before mixing, and again when cold. The mixture occupies less volume than the sum of the two volumes. For the best results the volume of the acid should be about three times that of the water.

93. Sulphuric Acid made on a Large Scale involves the same principles as shown in Experiment 55, excepting that SO_2 is obtained by burning S or roasting FeS_2 (pyrite),

Fig. 29.

and HNO_3 is made on the spot from $NaNO_3$ and H_2SO_4. SO_2 enters a large leaden chamber, often 100 to 300 feet long, and jets of steam and small portions of HNO_3 are also forced in. The "chamber acid" thus formed is very dilute, and must be evaporated first in leaden pans, and finally in glass or platinum retorts, since strong H_2SO_4, especially if hot, dissolves lead. See Experiment 124. Study Figure 29, and write the reactions. $2 HNO_3$ breaks up into $2 NO_2$, H_2O, and O.

SULPHURIC ACID.

94. Importance. — Sulphuric acid has been called, next to human food, the most indispensable article known. There is hardly a product of modern civilization in the manufacture of which it is not directly or indirectly used. Nearly a million tons are made yearly in Great Britain alone. It is the basis of all acids, as Na_2CO_3 is of alkalies. It is the life of chemical industry, and the quantity of it consumed is an index of a people's civilization. Only a few of its uses can be stated here. The two leading ones are the reduction of $Ca_3(PO_4)_2$ for artificial manures (see page 123) and the sodium carbonate manufacture. Foods depend on the productiveness of soils and on fertilizers, and thus indirectly our daily bread is supplied by means of this acid; and from sodium carbonate glass, soap, saleratus, baking-powders, and most alkalies are made directly or indirectly. H_2SO_4 is employed in bleaching, dyeing, printing, telegraphy, electroplating, galvanizing iron and wire, cleaning metals, refining Au and Ag, making alum, blacking, vitriols, glucose, mineral waters, ether, indigo, madder, nitro-glycerine, gun-cotton, parchment, celluloid, etc., etc.

FUMING SULPHURIC ACID.

95. Nordhausen or Fuming Sulphuric Acid, $H_2S_2O_7$, used in dissolving indigo and preparing coal-tar pigments, is made by distilling $FeSO_4$. $4\,FeSO_4 + H_2O = H_2S_2O_7 + 2\,Fe_2O_3 + 2\,SO_2$. This was the original sulphuric acid. It is also formed when SO_3 is dissolved in H_2SO_4. When exposed to the air, SO_3 escapes with fuming.

CHAPTER XX.

AMMONIUM HYDRATE.

96. Preparation of Bases. — We have seen that many acids are made by acting on a salt of the acid required, with a stronger acid. This is the direct way. The following experiments will show that bases may be prepared in a similar way by acting on salts of the base required with other bases, which we may regard as stronger than the ones to be obtained.

97. Preparation of NH_4OH and NH_3.

Experiment 57. — Powder 10^g ammonium chloride, NH_4Cl, in a mortar and mix with 10^g calcium hydrate, $Ca(OH)_2$; recently slaked lime is the best. Cover with water in a flask, and connect with Woulff bottles, as for making HCl (Fig. 22); heat the flask for fifteen minutes or more. The experiment may be tried on a smaller scale with a t.t. if desired.

The reaction is: $2 NH_4Cl + Ca(OH)_2 = CaCl_2 + 2 NH_4OH$. NH_4OH is broken up into NH_3, ammonia gas, and water. $NH_4OH = NH_3 + H_2O$. These pass over into the first bottle, where the water takes up the NH_3, for which it has great affinity. One volume of water at 0° will absorb more than 1000 volumes of NH_3. Thus NH_4OH may be called a solution of NH_3 in H_2O. Write the reaction.

Experiment 58. — Powder and mix 2 or 3^g each of ammonium nitrate, NH_4NO_3, and $Ca(OH)_2$; put them into a t.t., and heat slowly. Note the odor. $2 NH_4NO_3 + Ca(OH)_2 = ?$

98. Tests.

Experiment 59. — (1) Generate a little of the gas in a t.t., and note the odor. (2) Test the gas with wet red litmus paper. (3) Put a little HCl into an e.d., and pass over it the fumes of NH_3 from a d.t. Note the result, and write the equation. (4) Fill a small t.t. with the gas by upward displacement; then, while still inverted, put the mouth of the t.t. into water. Explain the rise of the water. (5) How might NH_4Cl be obtained from the NH_4OH in the Woulff bottles? (6) Test the liquid in each bottle with red litmus paper. (7) Add some from the first bottle to 5 or 10^{cc} of a solution of $FeSO_4$ or $FeCl_2$, and look for a ppt. State the reaction.

99. Formation. — Ammonia, hartshorn, exists in animal and vegetable compounds, in salts, and, in small quantities, in the atmosphere. Rain washes it from the atmosphere into the soil; plants take it from the soil; animals extract it from plants. Coal, bones, horns, etc., are the chief sources of it, and from them it is obtained by distillation. It results also from decomposing animal matter. NH_3 can be produced by the direct union of N and H, only by an electric discharge or by ozone (see page 85). It may be collected over Hg like other gases that are very soluble in water.

100. Uses. — Ammonium hydrate, NH_4OH, and ammonia, NH_3, are used in chemical operations, in making artificial ice, and to some extent in medicine; from them also may be obtained ammonium salts. State what you would put with NH_4OH to obtain $(NH_4)_2SO_4$. To obtain NH_4NO_3. The use of NH_4OH in the laboratory may be illustrated by the following experiment: —

Experiment 60. — Into a t.t. put 10^{cc} of a solution of ferrous sulphate, $FeSO_4$. Into another put 10^{cc} of sodium sulphate solution, Na_2SO_4. Add a little NH_4OH to each. Notice a ppt. in the one case but none in the other. If solutions of these two compounds were mixed, the metals Fe and Na could be separated by the addition of NH_4OH, similar to the separation of Ag and Cu by HCl. See page 58. Try the experiment.

CHAPTER XXI.

SODIUM HYDRATE.

101. Preparation.

Experiment 61. — Dissolve 3^g sodium carbonate, Na_2CO_3, in 10 or 15^{cc} H_2O in an e.d., and bring it to the boiling-point. Then add to this a mixture of 1 or 2^g calcium hydrate, $Ca(OH)_2$, in 5 or 10^{cc} H_2O. It will not dissolve. Boil the whole for five minutes. Then pour off the liquid which holds NaOH in solution. Evaporate if desired. This is the usual mode of preparing NaOH.

The reaction is $Na_2CO_3 + Ca(OH)_2 = 2\,NaOH + CaCO_3$. The residue is $Ca(OH)_2$ and $CaCO_3$; the solution contains NaOH, which can be solidified by evaporating the water. Sodium hydrate is an ingredient in the manufacture of hard soap, and for this use thousands of tons are made annually, mostly in Europe. It is an important laboratory reagent, its use being similar to that of ammonium hydrate. Exposed to the air, it takes up water and CO_2, forming a mixture of NaOH and Na_2CO_3. It is one of the strongest alkalies, and corrodes the skin.

Experiment 62. — Put 20^{cc} of H_2O in a receiver. With the forceps take a piece of Na, not larger than half a pea, from the naphtha in which it is kept, drop it into the H_2O, and at once cover the receiver loosely with paper or cardboard. Watch the action, as the Na decomposes H_2O. $HOH + Na = NaOH + H$. If the water be hot the action is so rapid that enough heat is produced to set the H on fire. That the gas is H can be shown by putting the Na under the mouth of a small inverted t.t., filled with cold water, in a water-pan. Na rises to the top, and the t.t. fills with H, which can be tested. NaOH dissolves in the water.

102. Properties.

Experiment 63. — (1) Test with red litmus paper the solutions obtained in the last two experiments. (2) To 5^{cc} of alum solution, $K_2Al_2(SO_4)_4$, add 2^{cc} of the liquid, and notice the color and form of the ppt.

POTASSIUM HYDRATE.

103. KOH is made in the Same Way as NaOH. — Describe the process in full (Experiment 61), and give the equation.

Experiment 64. — Drop a small piece of K into a receiver of H_2O, as in Experiment 62. The K must be very small, and the experiment should not be watched at too close a range. The receiver should not be covered with glass, but with paper. The H burns, uniting with O of the air. The purple color is imparted by the burning, or oxidation of small particles of K. Write the equation for the combustion of each.

H_2O might be considered the symbol of an acid, since it is the union of H and a negative element; or, if written HOH, it might be called a base, since it has a positive element and the (OH) radical. It is neutral to litmus, and on this account might be called a salt. It is better, however, to call it simply an oxide.

Potassium hydrate, caustic potash, is employed for the manufacture of soft soap. See page 187. As a chemical reagent its action is almost precisely like that of caustic soda, though it is usually considered a stronger base, as K is a more electro-positive element than Na.

CALCIUM HYDRATE.

104. Calcium Hydrate, the Most Common of the Bases, is nearly as important to them as H_2SO_4 is to acids. Since it is used to make the other bases, it might be called the strongest base, as H_2SO_4 is often called the strongest acid. The strength of an acid or base, however,

depends on the substance to which it is applied, as well as on itself, and for most purposes this one is classified as a weaker base than the three previously described.

Sulphuric acid, the most useful of the acids, is not made directly from its salts, but has to be synthesized. Calcium hydrate is also made by an indirect process, as follows:

$CaCO_3$, *i.e.* limestone, marble, etc., is burnt in kilns with C, a process which separates the gas, CO_2, according to the reaction: $CaCO_3 = CaO + CO_2$. CaO is unslaked lime, or quick-lime. On treating this with water, slaked lime, $Ca(OH)_2$ is formed, with generation of great heat. $CaO + H_2O = Ca(OH)_2$. Its affinity for H_2O is so great that it takes the latter from the air, if exposed.

Experiment 65. — Saturate some unslaked lime with water, in an e.d., and look for the results stated above, leaving it as long as may be necessary.

105. Résumé. — From the experiments in the last few chapters on the three divisions of chemical compounds, acids, bases and salts, we have seen (1) that acids and bases are the chemical opposites of each other; (2) that salts are formed by the union of acids and bases; (3) that some acids can be obtained from their salts by the action of a stronger acid; (4) that some bases can be got from salts by the similar action of other bases; (5) that the strongest acids and bases, as well as others, may be obtained in an indirect way by synthesis.

CHAPTER XXII.

OXIDES OF NITROGEN.

106. There are five oxides of N, only two of which are important.

NITROGEN MONOXIDE (N_2O).

107. Preparation.

Experiment 66. — Put into a flask, holding 200^{cc}, 10^g of ammonium nitrate, NH_4NO_3; heat it over wire gauze or asbestus in an iron plate, having a d.t. connected with a large t.t., which is held in a receiver of water, and from this t.t., another d.t. passing into a pneumatic trough, so as to collect the gas over water (Fig. 30). Have all the bearings tight. The reaction is $NH_4NO_3 = 2 H_2O + N_2O$. The t.t. is for collecting the H_2O.

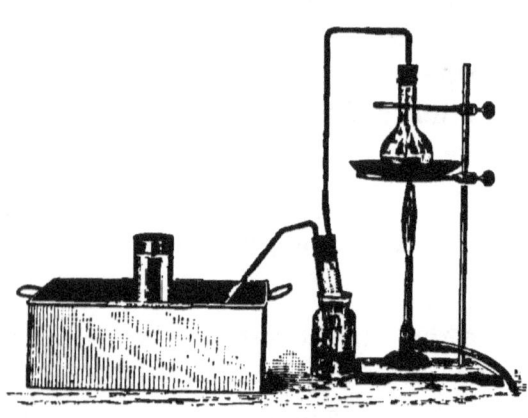

Fig. 30.

Note the color of the liquid in the t.t.; taste a drop, and test it with litmus. If the flask is heated too fast, some NO is formed, and this taking O from the air makes NO_2, which liquefies and gives an acid reaction and a red color. Some NH_4NO_3 is also liable to be carried over.

108. Properties.

Experiment 67. — Test the gas in the receiver with a burning stick and a glowing one, and compare the combustion with that in O. N_2O

may also be tested with S and P, if desired. N is set free in each case. Write the reactions.

Nitrogen monoxide or protoxide, the nitrous oxide of dentists, when inhaled, produces insensibility to pain, — anæsthesia, — and, if continued, death from suffocation. Birds die in half a minute from breathing it. Mixed with one-fourth O, and inhaled for a minute or two, it produces intoxication and laughter, and hence is called laughing gas. As made in Experiment 66, it contains Cl and NO, as impurities, and should not be breathed.

NITROGEN DIOXIDE (NO, or N_2O_2).

109. Preparation.

Experiment 68. — Into a t.t. or receiver put 5^g Cu turnings, add 5^{cc} H_2O and 5^{cc} HNO_3. Collect the gas like H, over water. $3\ Cu + 8\ HNO_3 = ?$ What two products will be left in the generator? Notice the color of the liquid. This color is characteristic of Cu salts. Notice also the red fumes of NO_2.

110. Properties.

Experiment 69. — Test the gas with a burning stick, admitting as little air as possible. Test it with burning S. NO is not a supporter of C and S combustion. Put a small bit of P in a deflagrating-spoon, and when it is vigorously burning, lower it into the gas. It should continue to burn. State the reaction. What combustion will NO support? Note that NO is half N, while N_2O is two-thirds N, and account for the difference in supporting combustion.

NITROGEN TETROXIDE (NO_2 or N_2O_4).

111. Preparation.

Experiment 70. — Lift from the water-pan a receiver of NO, and note the colored fumes. They are NO_2, or N_2O_4, nitrogen tetroxide. $NO + O = NO_2$. Is NO combustible? What is the source of O in the experiment?

NITROGEN TRIOXIDE (N_2O_3).

112. Preparation.

Experiment 71. — Put into a t.t. 1^g of starch and 1^{cc} of HNO_3. Heat the mixture for a minute. The red fumes are N_2O_3 and NO_2.

Nitrogen pentoxide, N_2O_5, is an unimportant solid. United with water it forms HNO_3. $N_2O_5 + H_2O = 2\ HNO_3$.

CHAPTER XXIII.

LAWS OF DEFINITE AND OF MULTIPLE PROPORTION.

113. Weight and Volume. — We have seen that water contains two parts of H by volume to one part of O; or, by weight, two parts of H to sixteen of O. These proportions are invariable, or no symbol for water would be possible. Every compound in the same way has an unvarying proportion of elements.

114. Law of Definite Proportion. — *In a given compound the proportion of any element by weight, or, if a gas, by volume is always constant.* Apply the law, by weight and by volume, to these: HCl, NH_3, H_2S, N_2O.

There is another law of equal importance in chemistry, which the compounds of N and O well illustrate.

		Weight.		Volume.	
		N.	O.	N.	O.
Nitrogen protoxide	N_2O,	28	16	2	1
Nitrogen dioxide	N_2O_2,	28	32	2	2
Nitrogen trioxide	N_2O_3,	28	48	2	3
Nitrogen tetroxide	N_2O_4,	28	64	2	4
Nitrogen pentoxide	N_2O_5,	28	80	2	5

Note that the proportion of O by weight is in each case a multiple of the first, 16. Also that the proportion by volume of O is a multiple of that in the first compound. In this example the N remains the same. If that had varied in the different compounds, it would also have

76 LAWS OF DEFINITE AND MULTIPLE PROPORTION.

varied by a multiple of the smallest proportion. This is true in all compounds.

115. Law of Multiple Proportion. — *Whenever one element combines with another in more than one proportion, it always combines in some multiple, one or more, of its least combining weight, or, if a gas, of its least combining volume.*

The least combining weight of an element is its atomic weight; and it is this fact of a least combining weight that leads us to believe the atom to be indivisible.

Apply the law in the case of P_2O, P_2O_3, P_2O_5; in $HClO$, $HClO_2$, $HClO_3$, $HClO_4$, arranging the symbols, weights, and volumes in a table, as above.

The volumetric proportions of each element in the oxides of nitrogen are exhibited below.

$$□ + □ + □ = \boxed{}$$
$$N + N + O = N_2O$$

$$□ + □ + □ + □ = \boxed{}$$
$$N + N + O + O = N_2O_2$$

$$□ + □ + □ + □ + □ = \boxed{}$$
$$N + N + O + O + O = N_2O_3$$

$$□ + □ + □ + □ + □ + □ = \boxed{}$$
$$N + N + O + O + O + O = N_2O_4$$

$$□ + □ + □ + □ + □ + □ + □ = \boxed{}$$
$$N + N + O + O + O + O + O = N_2O_5.$$

CHAPTER XXIV.

CARBON PROTOXIDE.

116. Preparation.

Experiment 72. — Put into a flask, of 200^{cc}, 5^g of oxalic acid crystals, $H_2C_2O_4$, and 25^{cc} H_2SO_4. Have the d.t. pass into a solution of NaOH in a Woulff bottle (Fig. 31), and collect the gas over water. Heat the flask slowly, and avoid inhaling the gas.

117. Tests.

Experiment 73. — Remove a receiver of the gas, and try to light the latter with a splinter. Is it combustible, or a supporter of (C) combustion? What is the color of the flame? When the combustion ceases, shake up a little lime water with the gas left in the receiver. What gas has been formed by the combustion, as shown by the test? See page 80. Give the reaction for the combustion.

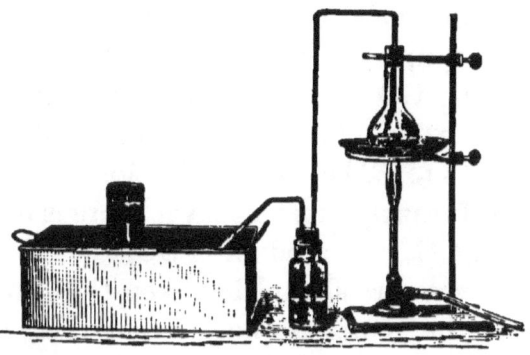

Fig. 31.

We have seen that H_2SO_4 has great affinity for H_2O. Oxalic acid consists of H, C, O in the right proportion to form H_2O, CO_2, and CO. H_2SO_4 withdraws H and O in the right proportion to form water, unites them, and then absorbs the water, leaving the C and O to combine and form CO_2 and CO. NaOH solution removes CO_2 from the mixture, forming Na_2CO_3, and leaves CO. Write both reactions.

118. Carbon Protoxide, called also carbon monoxide, carbonic oxide, etc., is a gas, having no color or taste, but possessing a faint odor. It is very poisonous. Being the lesser oxide of C, it is formed when C is burned in a limited supply of O, whereas CO_2 is always produced when O is abundant. The formation of each is well shown by tracing the combustion in a coal fire. Air enters at the bottom, and CO_2 is first formed. $C + 2 O = CO_2$. As this gas passes up, the white-hot coal removes one atom of O, leaving CO. $CO_2 + C = 2 CO$. At the top, if the draft be open, a blue flame shows the combustion of CO. $CO + O = CO_2$. The same reduction of CO_2 takes place in the iron furnace, and whenever there is not enough oxygen to form CO_2, the product is CO.

Great care should be taken that this gas does not escape into the room, as one per cent has proved fatal. Not all of it is burned at the top of the coal; and when the stove door is open, the upper drafts should be open also. It is the most poisonous of the gases from coal; hence the danger from sleeping in a room having a coal fire.

119. Water Gas. — CO is one of the constituents of "water gas," which, by reason of its cheapness, is supplanting gas made from coal, as an illuminator, in some cities. It is made by passing superheated steam over red-hot charcoal or coke. C unites with the O of H_2O, forming CO, and sets H free, thus producing two inflammable gases. $C + H_2O = ?$ As neither of these gives much light, naphtha is distilled and mixed with them in small quantities to furnish illuminating power See page 183.

CHAPTER XXV.

CARBON DIOXIDE.

120. Preparation.

Experiment 74. — Put into a t.t., or a bottle with a d.t. and a thistle-tube, 10 or 20^g $CaCO_3$, marble in lumps; add as many cubic centimeters of H_2O, and half as much HCl, and collect the gas by downward displacement (Fig. 39). Add more acid as needed. $CaCO_3 + 2\ HCl = CaCl_2 + H_2CO_3$. $H_2CO_3 = H_2O + CO_2$. H_2CO_3 is a very weak compound, and at once breaks up. By some, its existence as a compound is doubted.

121. Tests.

Experiment 75. — (1) Put a burning and a glowing stick into the t.t. or bottle. (2) Hold the end of the d.t. directly against the flame of a small burning stick. Does the gas support combustion? (3) Pour a receiver of the gas over a candle flame. What does this show of the weight of the gas? (4) Pass a little CO_2 into some H_2O (Fig. 32), and test it with litmus. Give the reaction for the solution of CO_2 in H_2O.

Experiment 76. — Put into a t.t. 5^{cc} of clear $Ca(OH)_2$ solution, *i.e.* lime water; insert in this the end of a d.t. from a CO_2 generator (Fig. 32). Notice any ppt. formed. It is $CaCO_3$. Let the action continue until the ppt. disappears and the liquid is clear. Then remove the d.t., boil the clear liquid for a minute, and notice whether the ppt. reappears.

122. Explanation.

$Ca(OH)_2 + CO_2 = CaCO_3 + H_2O$. The curious phenomena of this experiment are explained by the solubility of $CaCO_3$ in water containing CO_2, and its insolu-

Fig. 32.

bility in water, having no CO_2. When all the $Ca(OH)_2$ is combined, or changed to $CaCO_3$, the excess of CO_2 unites with H_2O, forming the weak acid H_2CO_3, which dissolves the precipitate, $CaCO_3$, and gives a clear liquid. On heating this, H_2CO_3 gives up its CO_2, and $CaCO_3$ is re-precipitated, not being soluble in pure water. See page 147.

Lime water, $Ca(OH)_2$ solution, is therefore a test for the presence of CO_2. To show that carbon dioxide is formed in breathing, and in the combustion of C, and that it is present in the air, perform the following experiment:

Experiment 77. — (1) Put a little lime water into a t.t., and blow into it through a piece of glass tubing. Any turbidity shows what? (2) Burn a candle for a few minutes in a receiver of air, then take out the candle and shake up lime water with the gas. (3) Expose some lime water in an e.d. to the air for some time.

123. Oxidation in the Human System. — Carbon dioxide, or carbonic anhydride, carbonic acid, etc., CO_2, is a heavy gas, without color or odor. It has a sharp, prickly taste, and is commonly reckoned as poisonous if inhaled in large quantities, though it does not chemically combine with the blood as CO does. Ten per cent in the air will sometimes produce death, and five per cent produces drowsiness. It exists in minute portions in the atmosphere, and often accumulates at the bottom of old wells and caverns, owing to its slow diffusive power. Before going down into one of these, the air should always be tested by lowering a lighted candle. If this is extinguished, there is danger. CO_2 is the deadly "choke damp" after a mine explosion, CH_4 being converted into CO_2 and H_2O; a great deal is liberated during volcanic eruptions, and it is formed in breathing by the union of O in the air with C in the system. This union of C and O takes place in the lungs and in all the tissues of the body, even on the surface. Oxygen is taken into the lungs, passes through the thin membrane into the blood, forms a weak chemical union with the red corpuscles, and is conveyed by them to all parts of the system. Throughout the body, wherever necessary, C and H are supplied for the O, and unite with it to form CO_2 and H_2O. These are taken up by the blood though they do not form a chemical union with it, are carried to the lungs, and pass out, together with the unused N and surplus O. The system is thus purified, and the waste must be supplied by food. The process also keeps up the heat of the body as really as the combustion of C or P in O produces heat. The temperature of the body does not vary much from 99° F., any excess of heat passing off through perspiration, and being changed into other forms of energy.

If, as in some fevers, the temperature rises above about 105° F., the blood corpuscles are killed, and the person dies. During violent exercise much material is consumed, circulation is rapid, and quick breathing ensues. Oxygen is necessary for life. A healthy person inhales plentifully; and this element is one of nature's best remedies for disease. Deep and continued inhalations in cold weather are better than furnace fires to heat the system. All animals breathe O and exhale CO_2. Fishes and other aquatic animals obtain it, not by decomposing H_2O, but from air dissolved in water. Being cold-blooded, they need relatively little; but if no fresh water is supplied to those in captivity, they soon die of O starvation.

124. Oxidation in Water. — Swift-running streams are clear and comparatively pure, because their organic impurities are constantly brought to the surface and oxidized, whereas in stagnant pools these impurities accumulate. Reservoirs of water for city supply have sometimes been freed from impurities by aëration, *i.e.* by forcing air into the water.

125. Deoxidation in Plants. — Since CO_2 is so constantly poured into the atmosphere, why does it not accumulate there in large quantity? Why is there not less free O in the air to-day than there was a thousand years ago? The answer to these questions is found in the growth of vegetation. In the leaf of every plant are thousands of little chemical laboratories; CO_2 diffused in small quantities in the air passes, together with a very little H_2O, into the leaf, usually from its under side, and is decomposed by the radiant energy of the sun. The C is built into the woody fiber of the tree, and the O is ready to be re-breathed or burned again. CO_2 contributes to the growth of plants, O to that of animals; and the constituents of the atmosphere vary little from one age to another. The compensation of nature is here well shown. Plants feed upon what animals discard, transforming it into material for the sustenance of the latter, while animals prepare food for plants. All the C in plants is supposed to come from the CO_2 in the atmosphere. Animals obtain their supply from plants. The utility of the small percentage of CO_2 in the air is thus seen.

126. Uses. — CO_2 is used in making "soda-water," and in chemical engines to put out fires in their early stages. In either case it may be prepared by treating Na_2CO_3 or $CaCO_3$ with H_2SO_4. Give the reactions. On a small scale CO_2 is made from $HNaCO_3$.

CARBON DIOXIDE.

CO_2 has a very weak affinity for water, but probably forms with it H_2CO_3. Much carbon dioxide can be forced into water under pressure. This forms soda-water, which really contains no soda. The justification for the name is the material from which it is sometimes made. Salts from H_2CO_3, called carbonates, are numerous, Na_2CO_3 and $CaCO_3$ being the most important.

CHAPTER XXVI.

OZONE.

127. Preparation.

Experiment 78. — Scrape off the oxide from the surface of a piece of phosphorus 2^{cm} long, put it into a wide-mouthed bottle, half cover the P with water, cover the bottle with a glass, and leave it for half an hour or more.

128. Tests.

Experiment 79. — Remove the glass cover, smell the gas, and hold in it some wet iodo-starch paper. See page 104. Look for any blue color. Iodine has been set free, according to the reaction, $2KI + O_3 = K_2O + O_2 + I_2$, and has imparted a blue color to the starch, and ordinary oxygen has been formed. Why will not oxygen set iodine free from KI? What besides ozone will liberate it? See page 104.

129. Ozone, oxidized oxygen, active oxygen, etc., is an allotropic form of O. Its molecule is O_3, while that of ordinary oxygen is O_2.

$$\underbrace{\square + \square + \square}_{\text{3 atoms oxygen}} = \underbrace{\square}_{\text{1 mol. ozone.}}$$

Three atoms of oxygen are condensed into the space of two atoms of ozone, or three molecules of O are condensed into two molecules of ozone, or three liters of O are condensed into two liters of ozone.

Ozone is thus formed by oxidizing ordinary oxygen. $O_2 + O = O_3$. This takes place during thunder storms and in artificial electrical discharges. The quantity of ozone produced is small, five per cent being the maximum, and the usual quantity is far less than that.

Ozone is a powerful oxidizing agent, and will change S, P, and As into their *ic* acids. Cotton cloth was formerly bleached, and linen is now bleached, by spreading it on the grass and leaving it for weeks to be acted on by ozone, which is usually present in the air in small quantities, especially in the country. Ozone is a disinfectant, like other bleaching agents, and serves to clear the air of noxious gases and germs of infectious diseases. So much ozone is reduced in this way that the air of cities contains less of it than country air. A third is consumed in uniting with the substance which it oxidizes, while two-thirds are changed into oxygen, as in Experiment 79.

It is unhealthful to breathe much ozone, but a little in the air is desirable for disinfection.

Ozone will cause the inert N of the air to unite with H, to form ammonia. No other agent capable of doing this is known, so that all the NH_3 in the air, in fact all ammonium compounds taken up by plants from soils and fertilizers, may have been made originally through the agency of ozone. At a low temperature ozone has been liquefied. It is then distinctly blue.

Electrolysis of water is the best mode of preparing this substance in quantity. When prepared from P it is mixed with P_2O_3.

CHAPTER XXVII.

CHEMISTRY OF THE ATMOSPHERE.

130. Constituents. — The four chief constituents of the atmosphere are N, O, H_2O, CO_2, in the order of their abundance. What experiments show the presence of N, O, and CO_2 in the air? See pages 22, 80. Set a pitcher of ice water in a warm room, and the moisture that collects on the outside is deposited from the air. This shows the presence of H_2O. Rain, clouds, fog, and dew prove the same. H_2SO_4 and $CaCl_2$, on exposure to air, take up water. Experiment 18 shows that there is not far from four times as much N as O by volume in air. Hence if the atmosphere were a compound of N and O, and the proportion of four to one were exact, its symbol would be N_4O.

131. Air not a Compound. — The following facts show that air is not a compound, but rather a mixture of these gases.

1. The proportion of N and O in the air, though it does not vary much, is not always exactly the same. This could not be true if it were a compound. Why?

2. If N_4O were dissolved in water, the N would be four times the O in volume; but when air is dissolved, less than twice as much N as O is taken up.

3. No heat or condensation takes place when four measures of N are brought in contact with one of O. It cannot then be N_4O, for the vapor density of N_4O would

be 36 — *i.e.* $(14 \times 4 + 16) \div 2$; but that of air is $14\frac{1}{2}$ nearly — *i.e.* $(14 \times 4 + 16) \div 5$. See page 108. Analysis shows about 79 parts of N to 21 parts of O by volume in air.

132. Water. — The volume of H_2O, watery vapor, in the atmosphere is very variable. Warm air will hold more than cold, and at any temperature air may be near saturation, *i.e.* having all it will hold at that temperature, or it may have little. But some is always present; though the hot desert winds of North Africa are not more than $\frac{1}{15}$ saturated. A cubic meter of air at 25°, when saturated, contains more than 22^g of water.

133. Carbon Dioxide. — Carbon dioxide does not make up more than three or four parts in ten thousand of the air; but, in the whole of the atmosphere, this gives a very large aggregate. Why does not CO_2 form a layer below the O and N? See page 114.

134. Other Ingredients. — Other substances are found in the air in minute portions, *e.g.* NH_3 constitutes nearly one-millionth. Air is also impregnated with living and dead germs, dust particles, unburned carbon, etc., but these for the most part are confined to the portion near the earth's surface. In pestilential regions the germs of disease are said sometimes to contaminate the air for miles around. See page 195.

CHAPTER XXVIII.

THE CHEMISTRY OF WATER.

135. Pure Water. — Review the experiments for electrolysis, and for burning H. Pure water is obtained by distillation.

Experiment 80. — Provide a glass tube 40 or 50^{cm} long and 3 or 4^{cm} in diameter. Fit to each end a cork with two perforations, through one of which a long tube passes the entire length of the larger tube (Fig. 32a). Connect one end of this with a flask of water arranged for heating; pass the other end into an open receptacle for collecting the distilled water. Into the other perforations lead short tubes, — the one for water to flow into the large tube from a jet; the other, for the same to flow out. This condenses the steam by circulating cold water around it. The apparatus is called a Liebig's condenser. Put water into the flask, boil it, and notice the condensed liquid. It is comparatively pure water; for most of the substances in solution have a higher boiling-point than water, and are left behind when it is vaporized.

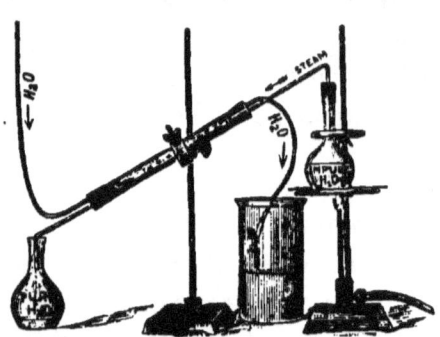

Fig. 32a.

136. Test.

Experiment 81. — Test the purity of distilled water by slowly evaporating a few drops on Pt foil in a room free from dust. There should be no spot or residue left on the foil. Test in the same way undistilled water.

THE CHEMISTRY OF WATER.

137. Water exists in Three States, — solid, liquid, and vaporous. It freezes at 0°, suddenly expanding considerably as it passes into the solid state. It boils, *i.e.* overcomes atmospheric pressure and is vaporized, at 100° (760^{mm} pressure). If the pressure is greater, the boiling-point is raised, *i.e.* it takes a higher temperature to overcome a greater pressure. If there be less pressure, as on a mountain, the boiling-point is lowered below 100°. Salts dissolved in water raise its boiling-point, and lower its freezing-point to an extent depending on the kind and quantity of the salt. Water, however, evaporates at all temperatures, even from ice.

Pure water has no taste or smell, and, in small quantities, no color. It is rarely if ever found on the earth. What is taken up by the air in evaporation is nearly pure; but when it falls as rain or snow, impurities are absorbed from the atmosphere. Water falling after a long rain, especially in the country, is tolerably free from impurities. Some springs have also nearly pure water; but to separate all foreign matter from it, water must be distilled. Even then it is liable to contain traces of ammonia, or some other substance which vaporizes at a lower temperature than water.

138. Sea-Water. — The ocean is the ultimate source of all water. From it and from lakes, rivers, and soils, water is taken into the atmosphere, falls as rain or snow, and sinks into the ground, reappearing in springs, or flowing off in brooks and rivers to the ocean or inland seas. Ocean water must naturally contain soluble salts; and many salts which are not soluble in pure water are dissolved in sea-water. In fact, there is a probability that all elements exist to some extent in sea-water, but many of them in extremely minute quantities. Sodium and magnesium salts are the two most abundant, and the bitter taste is due to $MgSO_4$ and $MgCl_2$. A liter of sea-water, nearly 1000^g, holds over 37^g of various salts, 29 of which are NaCl. See Hard Water, page 147.

139. River Water. — River water holds fewer salts, but has a great deal of organic matter, living and dead, derived from the regions through which it flows. To render this harmless for drinking, such water should be boiled, or filtered through unglazed porcelain. Carbon filters are now thought to possess but little virtue for separating harmful germs.

140. Spring Water. — The water of springs varies as widely in composition as do the rocks whence it bubbles forth. Sulphur springs contain much H_2S; many geysers hold SiO_2 in solution; chalybeate waters have compounds of Fe; others have Na_2SO_4, $MgSO_4$, NaCl, etc.

CHAPTER XXIX.

THE CHEMISTRY OF FLAME.

141. Candle Flame.

Experiment 82. — Examine a candle flame, holding a dark object behind it. Note three distinct portions: (1) a colorless interior about the wick, (2) a yellow light-giving portion beyond that, (3) a thin blue envelope outside of all, and scarcely discernible. Hold a small stick across the flame so that it may lie in all three parts, and observe that no combustion takes place in the inner portion.

142. Explanation. — A candle of paraffine, or tallow, is chiefly composed of compounds of C and H, in the solid state. The burning wick melts the solid; the liquid is then drawn up by the wick till the heat vaporizes and decomposes it, and O of the air comes in contact with the outer heated portion of gas, and burns it completely. Air tends to penetrate the whole body of the flame, but only N can pass through uncombined, for the O that is left after combustion in the outer portion seizes upon the compounds of C and H in the next, or yellow, part. There is not enough O here for complete combustion; at this temperature H burns before C, and the latter is set free. In that state it is of course a solid. Now an incandescent solid, or one glowing with heat, gives light, while the combustion of a gas gives scarcely any light, though it may produce great heat. While C in the middle flame is glowing, during the moment of its dissociation from H, it gives light. In the outer flame the temperature is high enough to burn entirely the gaseous compounds of C and H together, so that no solid C is set free, and hence no light is given except the faint blue. No combustion takes place in the inner blue cone, because no O reaches there.

By packing a wick into a cylindrical tin cup 5 or 10^{cm} high and 4^{cm} in diameter, containing alcohol, and lighting it, gunpowder can be held in the middle of the flame in a def. spoon, without burning. This shows the low temperature of that portion. Burning P will also be extinguished, thus showing the exclusion of O.

143. Bunsen Flame.

Experiment 83. — Examine a Bunsen burner. Unscrew the top, and note the orifices for the admission of gas and of air. Make a drawing. Replace the parts; then light the gas at the top, opening the air-holes at the base. Notice that the flame burns with very little color. Try to distinguish the three parts, as in the candle flame. These parts can best be seen by allowing direct sunlight to fall on the flame and observing its shadow on a white ground. Make a drawing of the flame. Hold across it a Pt wire and note at what part the wire glows most. Also press down on the flame for an instant with a cardboard or piece of paper; remove before it takes fire, and notice the charred circle. Put the end of a match into the blue cone, and note that it does not burn. Put the end of a Pt wire into this blue cone, and observe that it glows when near the top of the cone. What do these experiments show? Ascertain whether this inner portion contains a combustible material, by holding in it one end of a small d.t., and trying to ignite any gas escaping at the other end. It should burn. This shows that no combustion takes place in the interior of the flame, because sufficient free O is not present.

Next, close the air-holes, and note that the flame is yellow and gives much light. From this we infer the presence of solid particles in an incandescent state. But these could not come from the air. They must be C particles which have been set free from the C and H compounds of the gas, just as in the candle flame. The smoke that rises proves this. Hold an e.d. in the flame and collect some C. Try the same with the air-holes open.

144. Light and Heat of Flame. — Which of the two flames is hotter, the one with the air-holes open, or that with them closed? Evidently the former; for air is drawn in and mixes with the gas as it rises in the tube, and, on reaching the flame at the top, the two are well mingled, and the gaseous compounds of C and H burn at so high a temperature that solid C is not freed; hence there is little light. On closing the air-holes, no O can reach the flame except from the outside, and the heat is much less intense.

Fig. 33.

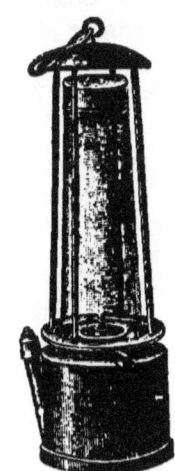
Fig. 34.

The H burns first, and sets the C free, which, while glowing, gives the light. This again illustrates the facts (1) that flame is caused by burning gas; (2) that light is produced by incandescent solids. Charcoal, coke, and anthracite coal burn without flame, or with very little, because of the absence of gases.

145. Temperature of Combustion.

Experiment 84. — Light a Bunsen flame, with the basal orifices open, and hold over it a fine wire gauze. Notice that the flame does not rise above the gauze. Extinguish the light, and try to ignite the gas above the gauze, holding the latter within 5 or 6^{cm} of the burner tube. Notice that it does not burn below the gauze (Fig. 33).

Gas and O are both present. Evidently, then, the only condition wanting for combustion is a sufficiently high temperature. The gauze cools the gas below its kindling-point.

This principle is made use of in the miner's lamp of Davy (Fig. 34). In coal mines a very inflammable gas, CH_4, called fire-damp, issues from the coal. If this collects in large quantities and mixes with O of the air, a kindling-point is all that is needed to make a violent explosion. An ordinary lamp would produce this, but the gauze lamp prevents it; for, though the inside may be filled with burning gas, CH_4, the flame cannot communicate with the outside.

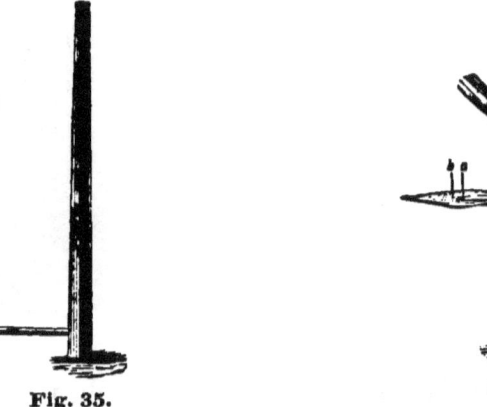

Fig. 35.

Fig. 36.
a, reducing flame. *b*, oxidizing flame.

146. Oxidizing and Reducing Flames. — The hottest part of a Bunsen flame is just above the inner blue cone (*b*, Fig. 36). Evidently there is more O at that point. If a reducing agent, *i.e.* a substance which takes up O, be put into this part of the flame, the latter will remove the O and appropriate it, forming an oxide. Cu heated there would become copper oxide. This part is called the oxidizing flame.

'The inner blue part of the Bunsen flame is devoid of O. It ought to remove O from an oxidizing agent, *i.e.* a substance which supplies O. If copper oxide be heated there (*a*, Fig. 36) by means of a mouth blow-pipe (Fig. 35), the flame will appropriate the O and leave the copper. This is called the reducing flame. Only the upper part of this blue central cone has heat enough to act in this way. By using a prepared piece of metal, to make the flame thin and to shut off the air, and then blowing the flame with a blow-pipe, greater strength can be obtained in both oxidizing and reducing flames (Fig. 36).

147. Combustible and Supporter Interchangeable.— H was found to burn in O. H was the combustible, O the supporter. Would O itself burn in H?—*i.e.* would the combustible become the supporter, and the supporter the combustible? As illuminating gas consists largely of H, and as air is part O, we may try the experiment with gas and air. Gas will burn in air. Will air burn in gas?

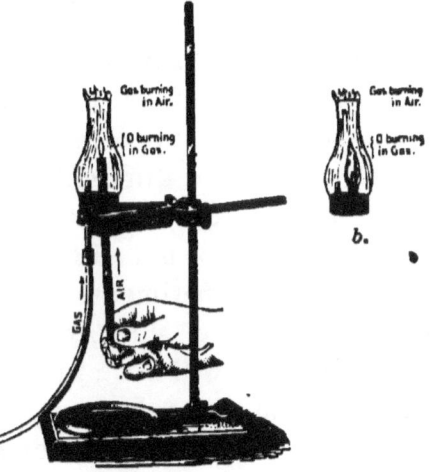

Fig. 37 *a*.

Experiment 85.— Fit a cork with two holes in it to the large end of a lamp chimney. Through each hole pass a short piece of tubing, and connect one of these with a rubber tube leading to a gas-jet. Pass a metallic tube, long enough to reach the top of the chimney, through the other, so that it will move easily up and down. Turn on the gas, and light it at the top of the chimney. Hold the end of the tube passing through the cork in the flame for a minute, then draw it down to the middle of the chimney (Fig. 37, *a*) and finally slowly remove it (*b*). Note that O from the air is burning in the gas. Which is the supporter, and which the combustible in this case? O will burn equally well in an atmosphere of H, as can be shown by experiment.

148. Explosive Mixture of Gases.

Experiment 86. — Slowly turn down the burning gas of a Bunsen lamp, having the orifices open, and notice that it suddenly explodes and goes out at the top, but now burns at the base. As the gas was gradually turned off, more air became mixed with it, until there was the right proportion of each gas for an explosion. Figure 38 shows the same thing. Light the gas at the top *a*, when the tube *c* covers the jet *b*. Then gradually raise the tube *c*. At a certain place there is the same explosion as with the lamp.

Fig. 38.

149. Generalizations. — These experiments show (1) that three conditions are necessary for combustion, — a combustible, a supporter, and a burning temperature which varies for different substances. Given these, "a fire" always results. The conditions for "spontaneous combustion" do not differ from those of any combustion. See Experiments 34, 112, 113, 114. (2) That combustible and supporter are interchangeable. If H burns in O, O will burn in H, the product being the same in each case. (3) For any combustion there must be a certain proportion of combustible and of supporter. Twenty per cent of CO_2 in the air dilutes the O to such an extent that C will not burn. Hence the utility of the chemical engine for putting out fires. (4) When two gases, a combustible and a supporter, are mixed in the requisite proportion, they form an explosive mixture, needing only the kindling temperature to unite them.

Chemical combination is always accompanied by disengagement of heat. Chemical dissociation is always accompanied by absorption of heat. The disengagement, or the absorption, is not always evident to the senses.

Combustion is the chemical combination of two or more substances with the self-evident disengagement of great heat, and usually of light.

The temperature of ignition varies greatly with different substances. PH_3 burns spontaneously at the usual temperatures of the air. P takes fire at 60°, but even at 10° it oxidizes with rapidity enough to produce phosphorescence. The vapor of CS_2 may be set on fire by a glass rod heated to 150°, but a red-hot iron will not ignite illuminating gas.

Spontaneous combustion often takes place in woolen or cotton rags which have been saturated with oil. The oil rapidly absorbs O, and sets fire to the cloth. This is thought to be the origin of some very destructive fires.

CHAPTER XXX.

CHLORINE.

150. Preparation.

Experiment 87.— Put into a t.t. 5ᵍ of fine granular MnO_2 and 10ᶜᶜ HCl. Apply heat carefully, and collect the gas by downward displacement in a receiver loosely covered with paper (Fig. 39). Add more HCl if needed. Have a good draft of air, and do not inhale the gas. If you have accidentally breathed it, inhale alcohol vapor from a handkerchief; alcohol has great affinity for Cl. Note the color of the gas, and compare its weight with that of air.

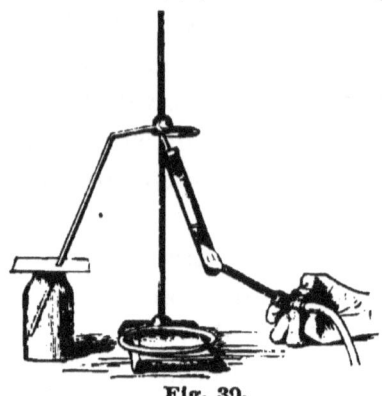

Fig. 39.

$MnO_2 + 4\ HCl = MnCl_2 + 2\ H_2O + 2\ Cl$. How much Cl can be separated with 5ᵍ MnO_2?

If preferred, a flask may be used for a generator instead of a t.t. Cl can be obtained directly from NaCl by adding H_2SO_4 (which produces HCl) and MnO_2. $2\ NaCl + 2\ H_2SO_4 + MnO_2 = MnSO_4 + Na_2SO_4 + 2\ H_2O + 2\ Cl$. Try the experiment, using a t.t. and adding water.

151. Cl from Bleaching-Powder.

Experiment 88.— Put a few grams of bleaching-powder into a small beaker, and set this into a larger one. Cover the latter with pasteboard or paper, through which passes a thistle-tube reaching into the small beaker (Fig. 40). Pour through the tube a little H_2SO_4 diluted with its volume of H_2O.

Fig. 40.

152. Chlorine Water.— A solution of Cl in water is often useful, and may be made as follows:—

CHLORINE 99

Experiment 89. — To 3 or 4 crystals of $KClO_3$ add a few drops of HCl. Heat a minute, and when the gas begins to disengage, pour in 10^{cc} H_2O, which dissolves the gas. $2 KClO_3 + 4 HCl = 2 KCl + Cl_2O_4 + 2 H_2O + 2 Cl$.

153. Bleaching Properties.

Experiment 90. — Put into a receiver of Cl, preferably before generating it, two pieces of Turkey red cloth, one wet, the other dry; a small piece of printed paper and a written one; also a red rose or a green leaf, each wet. Note from which the color is discharged. If it is not discharged from all, put a little H_2O into the receiver, shake it well, and state what ones are bleached.

Experiment 91. — (1) Add 5^{cc} of Cl water to 5^{cc} of indigo solution. (2) Treat in the same way 5^{cc} $K_2Cr_2O_7$ (potassium dichromate) solution, and record the results.

Indigo, writing-ink, and Turkey red or madder, are vegetable pigments; printer's ink contains C, and $K_2Cr_2O_7$ is a mineral pigment. State what coloring matters Cl will bleach.

154. Disinfecting Power.

Experiment 92. — Pass a little H_2S gas from a generator (page 120) into a t.t. containing Cl water. Look for a deposit of S. Notice that the odor of H_2S disappears. $H_2S + 2 Cl = 2 HCl + S$.

155. A Supporter of Combustion.

Experiment 93. — Sprinkle into a receiver of Cl a very little fine powder or filings of Cu, As, or Sb, and notice the combustion. Observe that here is a case of combustion in which O does not take part. Chlorides of the metals are of course formed. Write the reactions. See whether Cl will support the combustion of paper or of a stick of wood.

Experiment 94. — Warm 2 or 3^{cc} of oil of turpentine ($C_{10}H_{16}$) in an evaporating-dish; dip a piece of tissue paper into it, and very quickly thrust this into a receiver of Cl. It should take fire and deposit carbon. $C_{10}H_{16} + 16\,Cl = ?$ Test the moisture on the sides of the receiver with litmus. Clean the receiver with a little petroleum.

Experiment 95.— Prepare a H generator with a lamp-tube bent as in Figure 41. Light the H, observing the cautions in Experiment 23, and when well burning, lower the flame into a receiver of Cl. Observe the change of color which the flame undergoes as it comes in contact with Cl. Give the reaction for the burning. Test with litmus any moisture on the sides of the receiver. A mixture of Cl and H, in direct sunlight combines with explosive violence; whereas in diffused sunlight it combines slowly, and in darkness it does not combine. From these experiments state the chief properties of Cl, and what combustion it will support.

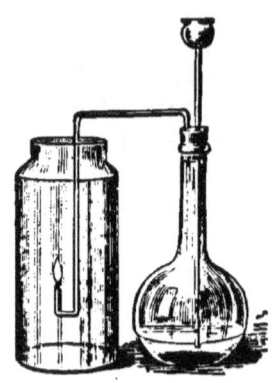

Fig. 41.

156. Sources and Uses. — The great source of Cl is NaCl, though it is often made from HCl. Its chief use is in making bleaching-powder, one pound of which will bleach 300 to 500 pounds of cloth. Cl is very easily liberated from this powder by a dilute acid, or, slowly, by taking moisture from the air. Hence its use as a disinfectant in destroying noxious gases and the germs of infectious diseases. Cl attacks organic matter and germs as it does the membrane of the throat or lungs, owing to its affinity for H.

Cl is the best bleaching agent for cotton goods. It is not suitable for animal materials, such as silk and wool, as it attacks their fiber. It does not discharge either mineral or carbon colors. The chemistry of bleaching is obscure. As dry material will not bleach, Cl seems to unite with H in H_2O and to set O free. The O then unites with some portion of the coloring matter, oxidizing it, and breaking up its molecule. Colors bleached by Cl cannot be restored.

CHAPTER XXXI.

BROMINE.

Examine bromine, potassium bromide, sodium bromide, magnesium bromide.

157. Preparation.

Experiment 96. — Pulverize 2 or 3g KBr, and mix it with about the same bulk of MnO_2. After putting this into a t.t., add as much H_2SO_4, mix them together by shaking, attach a d.t., and conduct the end of it into a t.t. that is immersed in a bottle of cold water. Slowly heat the contents of the t.t., and notice the color of the escaping vapor, and any liquid that condenses in the receiver. Avoid inhaling the fumes, or getting them into the eyes.

$MnO_2 + 2\,KBr + 2\,H_2SO_4 = ?$ Compare this with the equation for making Cl from NaCl.

158. Tests.

Experiment 97. — Try the bleaching action of Br vapor as in the case of Cl. Bleach a piece of litmus paper, and try to restore the color with NH_4OH. Explain its bleaching and disinfecting action. Try the combustibility of As, Sb, and Cu.

159. Description. — Bromine at usual temperatures is a liquid element; it is the only common one except Hg; it quickly evaporates on exposure to air. The chemistry of its manufacture is like that of Cl; its bleaching and disinfecting powers are similar to the latter, though they are not quite so strong as those of Cl. Its affinity for H and for metals is also strongly marked. A drop of Br on the skin produces a sore slow to heal.

Bromine salts are mainly KBr, NaBr, $MgBr_2$. These in small quantities accompany NaCl, and are most common in brine springs. The world's supply of Br comes chiefly from West Virginia and Ohio, over 300,000 pounds being produced from the salt (NaCl) wells there in 1884. The water taken from these wells is nearly evaporated, after which NaCl crystallizes out, leaving a thick liquid — bittern, or mother liquor — which contains the salts of Br. The bittern is treated with H_2SO_4 and MnO_2, as above.

For transportation in large quantities, Br has to be made into the salts NaBr and KBr, on account of the danger attending leakage or breakage of the receptacles for Br.

160. Uses. — Its chief uses are in photography (page 167), medicine, as KBr, and analytical chemistry.

CHAPTER XXXII.

IODINE.

Examine iodine, potassium iodide.

161. Preparation of I.

Experiment 98. — Put into a t.t. 2 or 3ᵍ of powdered KI mixed with an equal bulk of MnO_2, add H_2SO_4 enough to cover well, shake together, complete the apparatus as for making Br, and heat. Notice the color of the vapor, and any sublimate. The direct product of the solidification of a vapor is called a *sublimate.* The process is *sublimation.* Observe any crystals formed. Write the reaction, and compare the process with that for making Br and Cl. Compare the vapor density of I with that of Br and of Cl. With that of air. What vapor is heavier than I? See page 12. What acid and what base are represented by KI?

162. Tests.

Experiment 99. — (1) Put a crystal of I in the palm of the hand and watch it for a minute. (2) Put 2 or 3 crystals into a t.t., and warm it, meanwhile holding a stirring-rod half-way down the tube. Notice the vapor, also a sublimate on the sides of the t.t. and rod. (3) Add to 2 or 3 crystals in a t.t. 5ᶜᶜ of alcohol, C_2H_5OH; warm it, and see whether a solution is formed. If so, add 5ᶜᶜ H_2O and look for a ppt. of I. Does this show that I is not at all soluble in H_2O, or not so soluble as in alcohol?

163. Starch Solution and Iodine Test.

Experiment 100. — Pulverize a gram or two of starch, put it into an evaporating-dish, add 4 or 5 drops of water, and mix; then heat to the boiling-point 10ᶜᶜ H_2O in a t.t., and pour it over the starch, stirring it meanwhile.

(1) Dip into this starch paste a piece of paper, hold it in the vapor of I, and look for a change of color. (2) Pour a drop of the starch

paste into a clean t.t., and add a drop or two of the solution of I in alcohol. Add 5^{cc} H$_2$O, note the color, then boil, and finally cool. (3) The presence of starch in a potato or apple can be shown by putting a drop of I solution in alcohol on a slice of either, and observing the color. (4) Try to dissolve a few crystals of I in 5^{cc} H$_2$O by boiling. If it does not disappear, see whether any has dissolved, by touching a drop of the water to starch paste. This should show that I is slightly soluble in water.

164. Iodo-Starch Paper.

Experiment 101. — Add to some starch paste that contains no I 5^{cc} of a solution of KI, and stir the mixture. Why is it not colored blue? Dip into this several strips of paper, dry them, and save for use. This paper is called *iodo-starch paper*, and is used as a test for ozone, chlorine, etc. Bring a piece of it in contact with the vapor of chlorine, bromine, or ozone, and notice the blue color.

Experiment 102. — Add a few drops of chlorine water (see page 98) to 2^{cc} of the starch and KI solution in 10^{cc} H$_2$O. This should show the same effect as the previous experiment.

165. Explanation.
— Only free I, not compounds of it, will color starch blue. It must first be set free from KI. Ozone, chlorine, etc., have a strong affinity for K, and when brought in contact with KI they unite with K and set free I, which then acts on the starch present. Complete the equation: $KI + Cl = ?$

166. Occurrence.
— The ultimate source of I is sea water, of which it constitutes far too small a percentage to be separated artificially. Sea-weeds, or algæ, especially those growing in the deep sea, absorb its salts — NaI, KI, etc. — from the water. It thus forms a part of the plant, and from this much of the I of commerce is obtained. Algæ are collected in the spring, on the coasts of Ireland, Scotland, and Normandy, where rough weather throws them up. They are dried, and finally burned or distilled; the

ashes are leached to dissolve I salts; the water is nearly evaporated, and the residue is treated with H_2SO_4 and MnO_2, as in the case of Br and Cl. I also occurs in Chili, as NaI and $NaIO_3$, mixed with $NaNO_3$. This is an important source of the I supply.

167. Uses. — I is much used in medicine, and was formerly employed in taking daguerreotypes and photographs. Its solution in alcohol or in ether is known as tincture of iodine.

168. Fluorine. — F, Cl, Br, I, are called halogens or haloids, and exist in compounds — salts — in sea water. F is the most active of all elements, combining with every element except O. Until recently it has never been isolated, for as soon as set free from one compound it attacks the nearest substance, and seems to be as much averse to combining with itself, or to existing in the elementary state, as to uniting with O. It is supposed to be a gas, and, as is claimed, has lately been isolated by electrolysis from HF in a Pt U-tube. Fluorite (CaF_2) and cryolite ($Al_2F_6 + 6\,NaF$) are its two principal mineral sources. The enamel of the teeth contains F in composition.

CHAPTER XXXIII.

THE HALOGENS.

169. Halogens Compared. — The elements F, Cl, Br, I, form a natural group. Their properties, as well as those of their compounds, vary in a step-by-step way, as seen below. F is sometimes an exception. They are best remembered by comparing them with one another. Notice:

1. Similarity of name-ending. Each name ends in *ine*.
2. Similarity of origin. Salt water is the ultimate source of all, except F.
3. Similarity of valence. Each is usually a monad.
4. Similarity of preparation. Cl, Br, I, are obtained from their salts by means of MnO_2 and H_2SO_4.
5. Variation in occurrence. Cl occurs in sea-salt, Br in sea-water, I in sea-weed.
6. Variation in color; F being colorless, Cl green, Br red, I violet.
7. Gradation in sp. gr.; F 19, Cl 35.5, Br 80, I 127.
8. Gradation in state, corresponding to sp. gr.; F being a light gas, Cl a heavy gas, Br a liquid, I a solid.
9. Corresponding gradation in their usual chemical activity; F being most active, then Cl, Br, and I.
10. Corresponding gradation in the strength of the H acids; the strongest being HF, the next, HCl, etc.
11. Corresponding gradation in the explosibility of their N compounds; the strongest NCl_3, the next, NBr_3, etc.
12. Corresponding gradation in the number of H and O acids; Cl 4, Br 3, I 2.

170. Compounds. — The following are some of the oxides, acids, and salts of the halogens. Name them.

Cl_2O $(+H_2O=)$ 2 HClO. The salts are hypochlorites, as $Ca(ClO)_2$.
Cl_2O_3 $(+H_2O=)$ 2 HClO$_2$. The salts are chlorites, as $KClO_2$.
Cl_2O_4.
——— $HClO_3$. The salts are chlorates, as $KClO_3$.
——— $HClO_4$. The salts are perchlorates, as $KClO_4$.

——— HBrO. The salts are ? KBrO.
——— ——— The salts are wanting.
——— $HBrO_3$. The salts are ? $KBrO_3$.
——— $HBrO_4$. The salts are ? $KBrO_4$.

——— ——— The salts are wanting.
——— ——— The salts are wanting.
I_2O_5 $(+H_2O=)$ 2 HIO_3. The salts are ? KIO_3.
——— HIO_4. The salts are ? KIO_4.

F forms no oxides, and no acids except HF. HF, HCl, HBr, HI, are striking illustrations of acids with no O. $HClO_4$ is a very strong oxidizing agent. A drop of it will set paper on fire, or with powdered charcoal explode violently. This is owing to the ease with which it gives up O. Notice why its molecule is broken up more readily than $HClO_3$. The higher the molecular tower, or the more atoms it contains, the greater its liability to fall. Some organic compounds contain hundreds of atoms, and hence are easily broken down, or, as we say, are unstable. Inorganic compounds are, as a rule, much more stable than organic ones. It is not always true, however, that the compound with the least number of atoms is the most stable. SO_2 is more stable than SO_3, but H_2SO_3 is less so than H_2SO_4.

CHAPTER XXXIV.

VAPOR DENSITY AND MOLECULAR WEIGHT.

Examine a liter measure, in the form of a cube, — cubic decimeter, — and a cubic centimeter.

171. Gaseous Weights and Volumes. — A liter of H, at 0° and 760^{mm}, weighs nearly 0.09^g. This weight is called a crith. Find the weight of H in the following, in criths and in grams: 15^l, 0.07^l, 50.3^l, 0.035^l, 0.6^l.

It has been estimated that there are $(10)^{24}$ molecules of H in a liter. Does the number vary for different gases? The weight of a molecule of H in parts of a crith is $\frac{1}{(10)^{24}}$; in parts of a gram $\frac{.09}{(10)^{24}}$. If the H molecule is composed of 2 atoms, what is the weight of its atom in fractions of a crith? What in fractions of a gram? The weight of the H atom is a microcrith. What part of a crith is a microcrith?

172. Vapor Density. — Vapor density, or specific gravity referred to H as the standard (Physics, page 59), is the ratio of the weight of a given volume of a gas or vapor to the weight of the same volume of H. A liter of steam weighs nine times as much as a liter of H. Its vapor density is therefore nine. For convenience, a definite volume of H is usually taken as the standard, viz., the H atom. The volume of the H atom and that of the half-molecule of H_2O, or of any gas are identical, each being represented by one square, □. If, then, the standard of vapor density is the H atom, half the molecular weight of a gas must be

VAPOR DENSITY AND MOLECULAR WEIGHT.

its vapor density, since it is evident that we thus compare the weights of equal volumes. The vapor density of H_2O, steam, is found from the symbol as follows: $(2+16) \div 2 = 9$. To obtain the vapor density of any compound from the formula, we have only to divide its molecular weight by two. Find the vapor density of HCl, N_2O, NO, $C_{12}H_{22}O_{11}$, Cl, CO_2, HF, SO_2. Explain each case.

The half-molecule, instead of the whole, is taken, because our standard is the hydrogen atom □, the smallest portion of matter, by weight, known to science.

How many criths in a liter of HCl? How many grams? Compute the number of criths and of grams in one liter of the compounds whose symbols appear above.

PROBLEMS.

(1) A certain volume of H weighs 0.36g at standard temperature and pressure. How many liters does it contain? If one liter weighs 0.09g, to weigh 0.36g it will take $0.36 \div 0.09 = 4$ liters.

(2) How many liters, or criths, of H in 63g? 2.7g? 1g? 5g? 250g? Explain each.

(3) Suppose the gas to be twice as heavy as H, how many liters in 0.36g? A liter of the gas will weigh 0.18g (0.09×2). In 0.36g there will be $0.36 \div 0.18 = 2$. Answer the question for 63g, 2.7g, etc.

(4) How many liters of Cl in each of the above numbers of grams?

(5) How many of HCl? H_2O (steam)? CO_2? Explain fully every case.

Vapor density is very easily determined from the formula by the method given above. But in practice the formula is obtained from the vapor density, and hence the method there given has to be reversed.

173. Vapor Density of Oxygen. — Suppose we were to obtain the vapor density of O. We should carefully seal and weigh a given volume, say a liter, at a noted temperature and barometric pressure, which are reduced to 0° and 760^{mm}, and compare it with the weight of the same volume of H. This has been done repeatedly, and O has been found to weigh 16 times as much as H, volume for volume, or, more exactly, 15.96+. Now a liter of each gas has the same number of molecules, therefore the O molecule weighs 16 times the H molecule. The half-molecule of each has the same proportion, and the vapor density of O is 16. Atomic weight is obtained in a very different way.

PROBLEMS.

(1) A liter of Cl is found to weigh 3.195^g. Compute its vapor density, and explain fully.

(2) A liter of Hg vapor, under standard conditions, weighs 9^g. Find its vapor density, and explain.

The vapor density of only a few elements has been satisfactorily determined. See page 12. Some cannot be vaporized; others can be, but only under conditions which prevent weighing them. The vapor density of very many compounds also is unknown.

(3) A liter of CO_2 weighs 1.98^g. Find the vapor density, and from that the molecular weight, remembering that the latter is twice the former. See whether it corresponds to that obtained from the formula, CO_2. This is, in fact, the way a formula is ascertained, if the atomic weights of its elements are known.

(4) A liter of a compound gas weighs 2.88^g. Analysis shows that its weight is half S and half O. As the atomic weight of S is 32, and that of O is 16, what is the symbol for the gas?

Solution. Its molecular weight is 64, *i.e.* $(2.88 \div 0.09) \times 2$, of which 32 is S and 32 O. The atomic weight of S is 32, hence there is one atom of S, while of O there are two atoms. The formula is SO_2.

(5) A liter of a compound gas, which is found to contain $\frac{3}{4}$ C and $\frac{1}{4}$ O by weight, weighs 1.26^g. What is its formula? Atomic weights are taken from page 12. Prove your answer.

(6) A liter of a compound of N and O weighs 1.98^g. The N is $\frac{7}{11}$ and the O $\frac{4}{11}$. What is the gas?

(7) A compound of N and H gas weighs 0.765^g to the liter. The N is $\frac{14}{17}$ of the whole, the H $\frac{3}{17}$. What gas is it?

CHAPTER XXXV.

ATOMIC WEIGHT.

174. Definition. — We have seen that the molecular weight of a compound, as well as of most elements, is obtained from the vapor density by doubling the latter. It remains to explain how atomic weights are obtained. The term is rather misleading. The atomic weight of an element is its least combining weight, the smallest portion that enters into chemical union, which is, of course, the weight of an atom.

175. Atomic Weight of Oxygen. — Suppose we wish to find the atomic weight of oxygen. We must find the smallest proportion by weight in which it occurs in any compound. This can only be done by analyzing all the compounds of O that can be vaporized. As illustrative of these compounds take the six following: —

Names.	V.d.	Mol. wt.	Wt. of O.	Wt. of other Elem.	Symbol.
Carbon monoxide	14	28	16	12	?
Carbon dioxide	22	44	32	12	?
Hydrogen monoxide	9	18	16	2	?
Nitrogen monoxide	22	44	16	28	?
Nitrogen trioxide	38	76	48	28	?
Nitrogen pentoxide	54	108	80	28	?

176. Molecular Symbols. — From the vapor density of the gases — column 2 — we obtain their molecular weight — column 3. To find the proportion of O, it must be separated by chemical means from its compounds and separately weighed. These relative weights are given in column 4. Now the smallest weight of O which unites in any case is its atomic weight. If any compound of O should in future be found in which its combining weight is 8 or 4, that would be called its atomic weight. By dividing the numbers in column 4, wt. of O, by 16, the atomic weight of O, we obtain the number of O atoms in the molecule. Subtracting the weights of O from the molecular weights, we have the parts of the other elements, column 5, and dividing these by the atomic weight of the respective elements, we have the number of atoms of those elements; these last, combined with the number of O atoms, give the symbol. In this way complete the last column.

Show how to get the atomic weight of Cl from these compounds, arranging them in tabular form, and completing as above: HCl, KCl, NaCl, $ZnCl_2$, $MgCl_2$; the atomic weight of N in these: N_2O, NO, NH_3.

177. Molecular and Atomic Volumes. — We thus see that vapor density and atomic weight are obtained in two quite different ways. In the case of elements the two are usually identical, *i.e.* with the few whose vapor density is known; but this is not always true, and it leads to interesting conclusions regarding atomic volume. In O both vapor density and atomic weight are 16. This gives 2 atoms of O to the molecule, *i.e.* the molecular weight ÷ the atomic weight. The size of an O atom is therefore half the gaseous molecule, and is represented by one square, □. S has a vapor density and an atomic weight of 32 each. Compute the number of atoms in the molecule. Compute for I, in which the two are identical, 127. P has an atomic weight of 31, while its vapor density is 62. Its molecule must consist of 4 atoms, each half the size of the H atom,

☐. The vapor density of As is 150, the atomic weight 75. Compute the number of atoms in its molecule, and represent their relative size. Hg has an atomic weight of 200, a vapor density of 100. Compute as before, and compare the results with those on page 12. Ozone has an atomic weight of 16, a vapor density 24. Compute.

CHAPTER XXXVI.

DIFFUSION AND CONDENSATION OF GASES.

178. Diffusion of Gases. — Oxygen is 16 times as heavy as H. If the two gases were mixed, without combining, in a confined space, it might be supposed that O would settle to the bottom and H rise to the top. This would, in fact, take place at first, but only for an instant, for all gases tend to diffuse or become intimately mixed. The lighter the gas the more quickly it diffuses.

179. Law of Diffusion of Gases. — *The diffusibility of gases varies inversely as the square roots of their vapor densities.* Compare the diffusibility of H with that of O.

$$\text{dif. H} : \text{dif. O} :: \sqrt{16} : \sqrt{1}, \text{ or dif. H} : \text{dif. O} :: 4 : 1.$$

That is to say, if H and O be set free from separate receivers in a room, the H will become intermingled with the atmosphere four times as quickly as the O. Compare the diffusibility of O and N; of Cl and H. Take the atomic weights of these, since they are the same as the vapor densities. In case of a compound gas, half the molecular weight must be taken for the vapor density; *e.g.*

$$\text{dif. N}_2\text{O} : \text{dif. O} :: \sqrt{16} : \sqrt{22}.$$

180. Cause. — Diffusion is due to molecular motion; the lighter the gas the more rapid the vibration of its molecules. Compare the diffusibility of CO_2 and that of Cl; of HCl and SO_2; of HF and I.

181. Liquefaction and Solidification of Gases. — Water boils at 100°, under standard pressure, though evaporating at all temperatures; it vaporizes at a lower point if the pressure be less, as on a mountain, and at a higher temperature if the pressure be greater, as at points below the sea level. Alcohol boils at 78°, standard pressure, and every liquid has a point of temperature and pressure above which it must pass into the gaseous state. Likewise every gas has a *critical temperature* above which it cannot be liquefied at any pressure.

This condition was not recognized formerly, and before 1877, O, H, N, CH_4, CO, NO, etc., had not been liquefied, though put under a pressure of more than 2,000 atmospheres. They were called permanent gases. In 1877 Cailletet and Pictet liquefied and solidified these and others. The lowest temperature, about $-225°$, was produced by suddenly releasing the pressure from solid N to 4^{mm}, which caused it rapidly to evaporate. Evaporation, especially under diminished pressure, always lowers the temperature by withdrawing heat.

These low degrees are indicated by a H thermometer, or if too low for that, by a "thermo-electric couple" of copper and German silver.

The pupil can easily liquefy SO_2 by passing it through a U-tube which is surrounded by a mixture of ice and salt in a large receiver. At the meeting of the American Association for the Advancement of Science in 1887, a solid brick of CO_2 was seen and handled by the members. Liquid H is steel blue.

A few results obtained under a pressure of one atmosphere are: —

Boiling Points: $C_2H_4 - 102°$; $CH_4 - 184°$; $O - 181°$; $N - 194°$; $CO - 190°$; $NO - 154°$; Air $- 191°$.

Solidifying Points: $Cl - 102°$; $HCl - 115°$; Ether $- 129°$; Alcohol $- 130°$.

CHAPTER XXXVII.

SULPHUR.

Examine brimstone, flowers of sulphur, pyrite, chalcopyrite, sphalerite, galenite, gypsum, barite.

182. Separation.

Experiment 103. — To a solution of 2^g of sodium sulphide, Na_2S_2, in 10^{cc} H_2O add 3 or 4^{cc} HCl, and look for a ppt. Filter, and examine the residue. It is lac sulphur, or milk of sulphur.

183. Crystals from Fusion.

Experiment 104. — In a beaker of 25 or 50^{cc} capacity put 20^g brimstone. Place this over a flame with asbestos paper interposed, and melt it *slowly*. Note the color of the liquid, then let it cool, watching for crystals. When partly solidified pour the liquid portion into an evaporating-dish of water, and observe the crystals of S forming in the beaker (Fig. 42). The hard mass may be separated from the glass by a little HNO_3 and a thin knife-blade, or by CS_2.

Fig. 42.

184. Allotropy.

Experiment 105. — Place in a t.t. 15^g of brimstone, then heat slowly till it melts. Notice the thin amber-colored liquid. The temperature is now a little above 100°. As the heat increases, notice that it grows darker till it becomes black and so viscid that it cannot be poured out. It is now above 200°. Still heat, and observe that it changes to a slightly lighter color, and is again a thin liquid. At this time it is above 300°. Now pour a little into an evaporating dish containing water. Examine this, noticing that it can be stretched like rubber. Leave it in the water till it becomes hard. Continue heating the brimstone in the t.t. till it boils at about 450°, and note the color of the escaping vapor. Just above this point it takes fire. Cool the t.t., holding it in the light meantime, and look for a sublimate of S on the sides.

185. Solution.

Experiment 106. — Place in an evaporating-dish a gram of powdered brimstone, and add 5^{cc} CS_2, carbon disulphide. Stir, and see whether S is dissolved. Put this in a draft of air, and note the evaporation of the liquid CS_2 and the deposit of S crystals. These crystals are different in form from those resulting from cooling from fusion.

186. Theory of Allotropy. — The last three experiments well illustrate allotropy. We found S to crystallize in two different ways. Substances can crystallize in seven different systems, and usually a given substance is found in one of these systems only; *e.g.* galena is invariably cubical. An element having two such forms is said to be dimorphous. If it crystallizes in three systems, it is trimorphous. A crystal has a definite arrangement of its molecules. If without crystalline form, a substance is called amorphous. An illustration of amorphism was S after it had been poured into water. Thus S has at least three allotropic forms, and the gradations between these probably represent others. Allotropy seems to be due to varied molecular structure. We know but little of the molecular condition of solids and liquids, since we have no law to guide us like Avogadro's in gases; but, from the density of S vapor at different temperatures, we infer that liquids and solids have their molecules very differently made up from those of gases. The least combining weight of S is 32. Its vapor density at 1,000° is 32; hence its molecular weight is 64, *i.e.* vapor density $\times$ 2; and there are 2 atoms in its molecule at that temperature, molecular weight ÷ atomic weight. At 500°, however, the vapor density is 96 and the molecular weight 192. At this degree the molecule must contain 6 atoms. How many it has in the allotropic forms, as a solid, is beyond our knowledge; but it seems quite likely that allotropy is due to some change of molecular structure.

The above experiments show two modes of obtaining crystals, by fusion and by solution.

187. Occurrence and Purification. — Sulphur occurs both free and combined, and is a very common element. It is found free in all volcanic regions, but Sicily furnishes most of it. Great quantities are thrown up from the interior of the earth during an eruption. The heat of volcanic action probably separates it from its compound, which may be $CaSO_4$. Vast quantities of the poisonous SO_2 gas are also liberated during an eruption, this being, in volume of gases evolved, next to H_2O.

Fig. 43.

S is crudely separated from its earthy impurities in Sicily by piling it into heaps, covering to prevent access of air, and igniting, when some of the S burns, and the rest melts and is collected. After removal from the island it is further purified by distilling in retorts connected with large chambers where it sublimes on the sides as flowers of sulphur (Fig. 43). This is melted and run into molds, forming roll brimstone. S also occurs as a constituent of animal and vegetable compounds, as in mustard, hair, eggs, etc. The tarnishing of silver spoons by eggs is due to the formation of silver sulphide, Ag_2S. The yellow color of eggs, however, is due to oils, not to S.

The main compounds of S are sulphides and sulphates. What acids do they respectively represent? Metallic sulphides are as common as oxides; e.g. FeS_2, or pyrite, PbS, or galenite, ZnS, or sphalerite, $CuFeS_2$, or chalcopyrite, etc. The most abundant sulphate is $CaSO_4$, or gypsum. $BaSO_4$, or barite, and Na_2SO_4, or Glauber's salt, are others.

The only one of these compounds that is utilized for its S is FeS_2. In Europe this furnishes a great deal of the S for H_2SO_4. S is obtained by roasting FeS_2. $3 FeS_2 = Fe_3S_4 + 2 S$.

188. Uses. — The greatest use of S is in the manufacture of H_2SO_4. A great deal is used in making gunpowder, matches, vulcanized rubber, and the artificial sulphides, like HgS, H_2S, CS_2, etc. The last is a very volatile, ill-smelling liquid, made by the combination of two solids, S being passed over red-hot charcoal. It dissolves S, P, rubber, gums, and many other substances insoluble in H_2O.

189. Sulphur Dioxide, SO_2, has been made in many experiments. It is a bleaching agent, a disinfectant, and a very active compound, having great affinity for water, but it will not support combustion. Like most disinfectants, it is very injurious to the system. It is used to bleach silk and wool — animal substances — and straw goods, which Cl would injure; but the color can be restored, as the coloring molecule seems not to be broken up, but to combine with SO_2, which is again separated by reagents. Goods bleached with SO_2 often turn yellow after a time.

190. SO_2 a Bleacher.

Experiment 107. — Test its bleaching power by burning S under a receiver under which a wet rose or a green leaf is also placed.

CHAPTER XXXVIII.

HYDROGEN SULPHIDE.

Examine ferrous sulphide, natural and artificial.

191. Preparation.

Experiment 108. — Put a gram of ferrous sulphide (FeS) into a t.t. fitted with a d.t., as in Figure 32. Add 10^{cc} H_2O and 5^{cc} H_2SO_4. H_2S is formed. Write the equation, omitting H_2O. What is left in solution?

192. Tests.

Experiment 109. — (1) Take the odor of the escaping gas. (2) Pour into a t.t. 5^{cc} solution $AgNO_3$, and place the end of the d.t. from a H_2S generator into the solution and note the color of the ppt. What is the ppt.? Write the equation. (3) Experiment in the same way with $Pb(NO_3)_2$ solution. Write the equation. (4) Let some H_2S bubble into a t.t. of clean water. To see whether H_2S is soluble in H_2O, put a few drops of the water on a silver coin. Ag_2S is formed. Describe, and write the equation. Do the same with a copper coin. (5) Put a drop of lead acetate solution, $Pb(C_2H_3O_2)_2$, on a piece of unglazed paper, and hold this before the d.t. from which H_2S is escaping. PbS is formed. Write the equation. This is the characteristic test of H_2S.

193. Combustion of H_2S.

Experiment 110. — Attach a philosopher's lamp tube to the H_2S generator, and, observing the same precautions as with H, light the gas. What two products must be formed? State the reaction. The color of the flame. Compute the molecular weight and the vapor density of H_2S.

194. Uses. — Hydrogen sulphide or sulphuretted hydrogen, H_2S, is employed chiefly as a reagent in the chemical laboratory. It forms sulphides with many of the metals, as shown in the last experiment. These are precipitated from solution, and may be separated from other metals which are not so precipitated, as was found in the case of HCl and NH_4OH. The subjoined experiment will illustrate this. Suppose we wished to separate Pb from Ba, having salts of the two mixed together, as $Pb(NO_3)_2$ and $Ba(NO_3)_2$.

195. H_2S an Analyzer of Metals.

Experiment 111. — Pass some H_2S gas into 5^{cc} solution $Ba(NO_3)_2$. No ppt. is formed. Do the same with $Pb(NO_3)_2$ solution. A ppt. appears. Now mix 5^{cc} of each of these solutions in a t.t., and pass the gas from a H_2S generator into the liquid. What is precipitated, and what is unchanged? When fully saturated with the gas, as indicated by the smell, filter. Which metal is on the filter and which is in the filtrate? Other reagents, as Na_2CO_3 solution, would precipitate the latter.

196. Occurrence and Properties. — H_2S is an ill-smelling, poisonous gas, formed in sewers, rotten eggs, and other decaying albuminous matter. It is formed in the earth, probably from the action of water on sulphides, and issues with water from sulphur springs.

A characteristic property is the formation of metallic sulphides, as above. A skipper one night anchored his newly painted vessel near the Boston gas-house, where the refuse was deposited, with its escaping H_2S. In the morning, to his consternation, the craft was found to be black. H_2S had come in contact with the lead in the white paint, forming black PbS. This gradually oxidized after reaching the open sea, and the white color reappeared.

CHAPTER XXXIX.

PHOSPHORUS.

Note. — Phosphorus should be kept in water, and handled with forceps, never with the fingers, except under water, as it is liable to burn the flesh and produce ulcerating sores. Pieces not larger than half a pea should be used, and every bit should finally be burned.

197. Solution and Combustion.

Experiment 112. — Put 1 or 2 pieces of P into an evaporating-dish, and pour over them 5 or 10^{cc} CS_2, — carbon disulphide. This will be enough for a class. When dissolved, dip pieces of unglazed paper into it, and hold these in the air, looking for any combustion as they dry. The P is finely divided in solution, which accounts for its more ready combustion then. Notice that the paper is not destroyed. This is an example of so-called "spontaneous combustion." The burning-point of P, the combustible, in air, the supporter, is about $60°$.

198. Combustion under Water.

Experiment 113. — Put a piece of P in a t.t. which rests in a receiver, add a few crystals $KClO_3$ and 5^{cc} H_2O. Now pour in through a thistle-tube 1^{cc} or more of H_2SO_4. Look for any flame. H_2SO_4 acts very strongly on $KClO_3$. What is set free? From this fact explain the combustion in water.

199. Occurrence.

— P is very widely disseminated, but not abundant, and is found only in compounds, the chief of which is calcium phosphate $Ca_3(PO_4)_2$. It occurs in granite and other rocks, as the mineral apatite, in soils, in plants, particularly in seeds and grains, and in the bones, brains, etc., of vertebrates. From the human system it is excreted by the kidneys as microcosmic salt, $HNaNH_4PO_4$;

and when the brain is hard-worked, more than usual is excreted. Hence brain-workers have been said to "burn phosphorus."

200. Sources. — Rocks are the ultimate source of this element. These, by the action of heat, rain, and frost, are disintegrated and go to make soils. The rootlets of plants are sent through the soil, and, among other things, soluble phosphates in the earth are absorbed, circulated by the sap, and selected by the various tissues. Animals feed on plants, and the phosphates are circulated through the blood, and deposited in the osseous tissue, or wherever needed.

Human bones contain nearly 60 per cent of $Ca_3(PO_4)_2$; those of some birds over 80 per cent.

The main sources of phosphates and P are the phosphate beds of South Carolina, the apatite beds of Canada, and the bones of animals.

201. Preparation of Phosphates and Phosphorus. — Bone ash, obtained by burning or distilling bones, and grinding the residue, is treated with H_2SO_4, and forms soluble $H_4Ca(PO_4)_2$, superphosphate of lime, and insoluble $CaSO_4$.

$$Ca_3(PO_4)_2 + 2\ H_2SO_4 = H_4Ca(PO_4)_2 + 2\ CaSO_4.$$

This completes the process for fertilizers. If P is desired, the above is filtered; charcoal, a reducing agent, is added to the filtrate; the substance is evaporated, then very strongly heated and distilled in retorts, the necks of which dip under water. It is then purified from any uncombined C by melting in hot water and passing into molds in cold water.

The work is very dangerous and injurious, on account of the low burning-point of P, and its poisonous properties. While its compounds are necessary to human life, P itself destroys the bones, particularly the jaw bones, of the workers in it.

Between 1,000 and 2,000 tons are made yearly, mostly for matches, but almost all at two factories, one in England, and one in France.

202. Properties. — P is a colorless, transparent solid, when pure; the impure article is yellowish, translucent, and waxy. It is insoluble in water, slightly soluble in alcohol and ether, and it readily dissolves in CS_2, oil of turpentine, etc. Fumes, having a garlic odor, rise when it is exposed to the air, and in the dark it is phosphorescent, emitting a greenish light.

203. Uses. — The uses of this element and its compounds are for fertilizers, matches, vermin poisons, and chemical operations.

204. Matches. — The use of P for matches depends on its low burning-point. Prepared wood is dipped into melted S, and the end is then pressed against a stone slab having on it a paste of P, $KClO_3$, and glue. KNO_3 is often used instead of $KClO_3$. In either case the object is to furnish O to burn P. Matches containing $KClO_3$ snap on being scratched, while those having KNO_3 burn quietly. The friction from scratching a match generates heat enough to ignite the P, that enough to set the S on fire, and the S enough to burn the wood. Give the reaction for each. Paraffine is much used instead of S. Safety matches have no P, and must be scratched on a surface of red P and Sb_2S_3, or on glass.

205. Red Phosphorus. — Two or three allotropic forms of P are known, the principal one being red. If heated between 230° and 260°, away from air, the yellow variety changes to red, which can be kept at all temperatures below 260°. Above that it changes back. Red P is not poisonous, ignites only at a high temperature, and is not phosphorescent, like the yellow.

206. Spontaneous Combustion of Phosphene, or Hydrogen Phosphide, PH_3.

Experiment 114.— Put into a 200^{cc} flask 1^g P and 50^{cc} saturated solution NaOH or KOH. Connect with the p.t. by a long d.t., as in Figure 44, the end of which must be kept under water. Pour 3 or 4^{cc} of ether into the flask, to drive out the air. It is necessary to exclude all air, as a dangerously explosive mixture is formed with it. Heat the

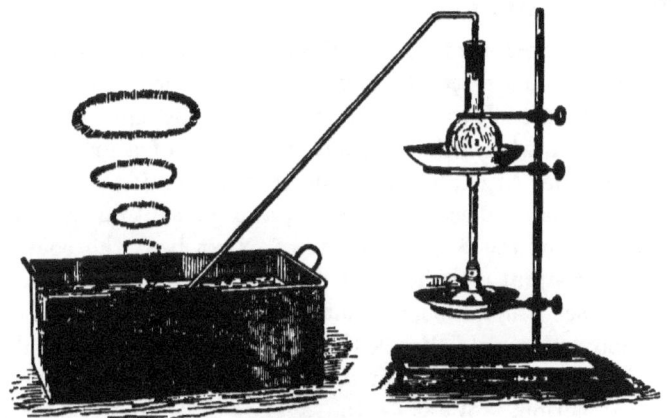

Fig. 44.

mixture, and as the gas passes over and into the air, it takes fire spontaneously, and rings of smoke successively rise. It will do no harm if, on taking away the lamp, the water is drawn back into the flask; but in that case the flask should be slightly lifted to prevent breakage by the sudden rush of water. On no account let the air be drawn over. The experiment has no practical value, but is an interesting illustration of the spontaneous combustion of PH_3 and of vortex rings. What are the products of the combustion? An admixture of another compound of P and H causes the combustion.

CHAPTER XL.

ARSENIC.

Examine metallic arsenic, realgar, orpiment, arsenopyrite, arsenic trioxide, copper arsenite.

The compounds of arsenic are very poisonous if taken into the system, and must be handled with care.

207. Separation.

Experiment 115. — Draw out into two parts in the Bunsen flame a piece of glass tubing 20^{cm} long and 1 or 2^{cm} in diameter. Into the end of one of the ignition tubes thus formed, when it is cool, put one-fourth of a gram of arsenic trioxide, As_2O_3, using paper to transfer it. Now put into the tube a piece of charcoal, and press it down to within 2 or 3^{cm} of the As_2O_3 (Fig. 45). Next heat the coal red-hot, and then at once heat the As_2O_3. Continue this process till you see a metallic sublimate — metallic mirror — on the tube above the coal. Break the tube and examine the sublimate. It is As. Heat vaporizes the As_2O_3. Explain the chemical action. What is the agency of C in the experiment? Of As_2O_3? $2 As_2O_3 + 3 C = ?$

Fig. 45.

208. Tests. — Experiments 115 and 116 are used as tests for the presence of arsenic.

Experiment 116. — Prepare a H generator, — a flask with a thistle-tube and a philosopher's lamp tube (Fig. 46), put in some granulated Zn, water, and HCl. Test the purity of the escaping gas (Experiment 23), and when pure, light the jet of H. H is now burning in air. To be sure that there is no As in the ingredients used, hold the inside of a porcelain evaporating-dish directly against the flame for a minute. If no silvery-white mirror is found, the chemicals are free from As. Then pour through the thistle-tube, while the lamp is still burning, 1^{cc} solution of As_2O_3 in HCl or H_2O — a bit of As_2O_3 not larger than a grain of wheat in 10^{cc} HCl. See whether the color of the flame changes; then hold the evaporating-dish once more in the flame, and notice a metallic deposit of As. Set away the apparatus under the hood and leave the light burning.

This experiment must not be performed unless all the cautions are observed, since the gas in the flask (AsH_3) is the most poisonous known, and a single bubble of it inhaled is said to have killed the discoverer. By confining the gas inside the flask there is no danger.

Fig. 46.

Instead of using As_2O_3 solution, a little Paris green, wall paper suspected of containing arsenic, green silk, or green paper labels, etc., may be soaked in HCl, and tested.

209. Explanation. — The chemical changes are as follows: The compounds of As, in this case As_2O_3, in presence of nascent H, are immediately converted into the deadly hydrogen arsenide (arsine, arseniuretted hydrogen), AsH_3. $As_2O_3 + 12\,H = 2\,AsH_3 + 3\,H_2O$. The AsH_3 mixed with excess of H tends to escape and is burned to As_2O_3 and H_2O, and thus is rendered comparatively harmless as it passes into the air. This is why the flame must be burning when the arsenic compound is introduced. $2\,AsH_3 + 6\,O = As_2O_3 + 3\,H_2O$.

In the combustion of AsH_3, H burns at a lower point than As. The introduction of a cold body like porcelain cools the flame below the kindling-point of As, and this is deposited, while H burns, in exactly the same way as lamp-black was collected in Experiment 26.

210. Expert Analysis. — A modification of this experiment is employed by experts to test for As_2O_3 poisoning.

The organs—stomach or liver—are cut into small pieces, dissolved by nascent Cl, or HClO, made from $KClO_3$ and HCl, and the solution is introduced into a H generator, as above. As_2O_3 preserves the tissues it comes in contact with, for a long time, and the test can be made years after death. All the chemicals must be pure, since As is found in small quantities in most ores, and the Zn, HCl, and H_2SO_4 of commerce are very likely to contain it. The above is called Marsh's test, and is so delicate that a mere trace of arsenic can be detected.

211. Properties and Occurrence. — As is a grayish white solid, of metallic luster, while a few of its characters are non-metallic. It is very widely distributed, being sometimes found native, and sometimes combined, as AsS, realgar, As_2S_3, orpiment, and FeAsS, arsenopyrite. Its chief source is the last, the fine powder of which is strongly heated, when As separates and sublimes. It has the odor of garlic, as may be observed by heating a little on charcoal with the blow-pipe.

212. Atomic Volume. — As is peculiar in that its atomic volume, so far as the volume can be determined, is only half that of the H atom. Its vapor density is 150, which gives 300 for the molecular weight, while its least combining or atomic weight is 75. 300, the molecular weight, $\div 75$, the atomic weight, $= 4$, the number of atoms in the molecule. All gaseous molecules being of the same size, represented by two squares, the atomic volume of As must be one-fourth of this size, represented by half of one square, ▫. Of what other element is this true? See page 12.

213. Uses of As_2O_3. — Arsenic is used in shot-manufacture, for hardening the metal. Its most important compound is As_2O_3, arsenic trioxide, called also arsenious anhydride, arsenious acid, white arsenic, etc. So poisonous is this that enough could be piled on a one-cent piece to kill a dozen persons. Taken in too large quantities it acts as an emetic. The antidote is ferric hydrate $Fe_2(OH)_6$ and a mustard emetic, followed by oil or milk.

The vapor density of this compound shows that its symbol should be As_4O_6, but the improper one, As_2O_3, is likely to remain in use. Another oxide, As_2O_5, arsenic pentoxide, exists, but is less important. Show how the respective acid formulæ are obtained from these anhydrides. See page 50.

As_2O_3 is used in making Paris green (page 165); in many green coloring materials, in which it exists as copper arsenite; in coloring wall papers, and in fly and rat poisons. It is employed for preserving skins, etc. Fashionable women sometimes eat it for the purpose of beautifying the complexion, to which it imparts a ghastly white, unhealthy hue. Mountaineers in some parts of Europe eat it for the greater power of endurance which it is supposed to give them. By beginning with small doses these arsenic-eaters finally consume a considerable quantity of the poison with apparent impunity; but as soon as the habit is stopped, all the pangs of arsenic-poisoning set in. Wall paper containing arsenic is said to be injurious to some people, while apparently harmless to others.

CHAPTER XLI.

SILICON, SILICA, AND SILICATES.

214. Comparison of Si and C. — The element Si resembles carbon in valence and in allotropic forms. It occurs in three forms like C, — a diamond form, a graphite, and an amorphous. C forms the basis of the vegetable and animal world; Si, of the mineral. Most soils and rocks, except limestone, are mainly compounds of O, Si, and metals. While O is estimated to make up nearly one-half of the known crust of the earth, Si constitutes fully a third. The two are usually combined, as silica, SiO_2, or silicates, SiO_2 combined with metallic oxides. This affinity for O is so strong that Si is not found uncombined, and is separated with great difficulty and only at the highest temperatures. No special use has yet been found for it, except as an alloy with Al. Its compounds are very important.

215 Silica. — Examine some specimens of quartz, rock crystal, white and colored sands, agate, jasper, flint, etc.; test their hardness with a knife blade, and see whether they will scratch glass. Notice that quartz crystals are hexagonal or six-sided prisms, terminated by hexagonal pyramids. The coloring matters are impurities, often Fe and Mn, if red or brown. When pure, quartz is transparent as glass, infusible except in the oxy-hydrogen blow-pipe, and harder than glass. Rock crystal is massive SiO_2. Sand is generally either silica or silicates.

The common variety of SiO_2 is not soluble in water or in acids, except HF. An amorphous variety is to some extent soluble in water. Most geysers deposit the latter in successive layers about their mouths. Agate, chalcedony, and opal have probably an origin similar to this. A solution of this variety of SiO_2 forms a jelly-like mass — colloid — which will not diffuse through a membrane of parchment — dialyzer — when suspended in water. Crystalloids will diffuse through such a membrane, if they are in solution. This principle forms the basis of dialysis.

All substances are supposed to be either crystalloids, *i.e.* susceptible of crystallization, or colloids — jelly-like masses. HCl is the most diffusible in liquids of all known substances; caramel is one of the least so. To separate the two, they would be put into a dialyzer suspended in water, when HCl will diffuse through into the water, and caramel will remain. As_2O_3, in cases of suspected poisoning, was formerly separated from the stomach in this way, as it is a crystalloid, whereas most of the other contents of the stomach are colloidal.

216. Silicates. — Si is a tetrad (page 12). $SiO_2 + 2 H_2O = ?$ $SiO_2 + H_2O = ?$ In either case the product is called silicic acid. Replace all the H with Na, and name the product. Replace it with K; Mg; Fe; Pb; Ca. Na_4SiO_4 and Na_2SiO_3 are typical silicates of Na, but others exist.

217. Formation of SiO_2 from Sodium Silicate.

Experiment 117. — To 5^{cc} Na_4SiO_4 in an evaporating-dish add 5^{cc} HCl. Describe the effect. Pour away any extra HCl. Heat the residue gently, above a flame, till it becomes white, then cool it and add water. In a few minutes taste a drop of the water, then pour it off, leaving the residue. Crush a little in the fingers, and compare it with white sand, SiO_2. Apply to the experiment these equations : —

$Na_4SiO_4 + 4 HCl = 4 NaCl + H_4SiO_4$. $H_4SiO_4 - 2 H_2O = SiO_2$. Why was H_4SiO_4 heated? Why was water finally added?

Water glass, sodium or potassium silicate, used somewhat for making artificial stone, is made by fusing SiO_2 with Na_2CO_3 or K_2CO_3, and dissolving in water. Silicic acid forms the basis of a very important series of compounds, — the silicates. The above two are the only soluble ones, and may be called liquid glass. See page 172.

CHAPTER XLII.

GLASS AND POTTERY.

Examine white sand, calcium carbonate, sodium carbonate, smalt; bottle, window, Bohemian and flint glass.

218. Glass is an Artificial Silicate. — SiO_2 alone is almost infusible, as is also CaO; but mixed and heated the two readily fuse, forming calcium silicate. $CaO + SiO_2 = ?$ Notice that SiO_2 is the basis of an acid, while CaO is essentially a base, and the union of the two forms a salt. There are four principal kinds of glass: (1) Bohemian, a silicate of K and Ca, not easily fused, and hence used for chemical apparatus where high temperatures are required; (2) window or plate glass, a silicate of Na and Ca; (3) bottle glass, a silicate of Na, Ca, Al, Fe, etc., a variety which is impure, and is tinged green by salts of Fe; (4) flint glass, a silicate of K and Pb, used for lenses in optical instruments, cut glass ware, and, with B added, for paste, or imitation diamonds, etc. Pb gives to glass high refracting power, which is a valuable property of diamonds, as well as of lenses.

219. Manufacture. — Pure white sand, SiO_2, is mixed with $CaCO_3$ and Na_2CO_3, some old glass — cullet — is added, and the mixture is fused in fire-clay crucibles. For flint glass, Pb_3O_4, red lead, is employed. If color is desired, mineral coloring matter is also added, but not always at this stage. CoO, or smalt, gives blue; uranium oxide, green; a mixture of Au and Sn of uncertain composition, called the "purple of Cassius," gives purple. MnO_2 is used to correct the green tint caused by FeO, which it is supposed to oxidize. Opacity, or enamel, as in lamp-shades, is produced by adding As_2O_3, Sb_2O_3, SnO_2, cryolite, etc. The glass-worker dips his blow-pipe — a hollow iron rod five or six feet long — into the fused mass of glass, removes a small portion, rolls it on a smooth surface, swings it round in the air, blowing meanwhile through the rod, and thus fashions it as desired, into bottles, flasks, etc. For some wares, *e.g.* common goblets, the glass is run into molds and stamped; for others it is blown and welded. All glass must be annealed, *i.e.* cooled slowly, for several days. The molecules thus arrange themselves naturally. If not annealed, it breaks very easily. It may be greatly toughened by dipping, when nearly red-hot, into hot oil. Cut glass is prepared at great expense by subsequent grinding. Glass may be rendered semi-opaque by etching either with HF, or with a blast of sand.

220. Importance. — Few manufactured articles have more importance than glass. Without it the sciences of chemistry, physics, astronomy, microscopic anatomy, zoölogy, and botany, not to mention its domestic uses, would be almost impossible.

221. Porcelain and Pottery. — Genuine porcelain and china-ware are made of a fine clay, kaolin, which results from the disintegration of feldspathic rocks. Bricks are baked clay. The FeO in common clay is oxidized to Fe_2O_3, on heating, a process which gives their red color. Some clay, having no Fe, is white; this is used for fire-bricks and clay pipes. That containing Fe is too fusible for fire-clay, which must also have much SiO_2. The electric arc, however, will melt even this, and the most refractory vessels are of calcium oxide or of graphite.

Pottery is clay, molded, baked, and either glazed, like crockery, or unglazed, like flower-pots. Jugs and coarse earthenware are glazed by volatilizing NaCl in an oven which holds the porous material. This coats the ware with sodium silicate. To glaze china, it is dipped into a powder of feldspar and SiO_2 suspended in water and vinegar, and then fused. If the ware and glaze expand uniformly with heat, the latter does not crack.

CHAPTER XLIII.

METALS AND THEIR ALLOYS.

222. Comparison of Metals and Non-Metals. — The majority of elements are metals, only about a dozen being non-metallic in their properties. The division line between the two classes is not very well defined; *e.g.* As has certain properties which ally it to metals; it has other properties which are non-metallic. H occupies a place between the two classes. The following are the more marked characteristics of each group: —

METALS.	NON-METALS.
1. Metals are solid at ordinary temperatures, and usually of high specific gravity. Exceptions: Hg is liquid above $-39.5°$; Li is the lightest solid known; Na and K will float on water.	1. Non-metals are either gaseous or solid at ordinary temperatures, and of low specific gravity. Exceptions: Br is a liquid; I has the heaviest known vapor.
2. Metals reflect light in a way peculiar to themselves. They have what is called a metallic luster.	2. Non-metallic solids have different lusters, as glassy, resinous silky, etc. Exceptions: I, B, and C have metallic luster.
3. They are white or gray. Exceptions: Au, Ca, Sr are yellow; Cu is red.	3. Non-metals have no characteristic color.
4. In general they conduct heat and electricity well.	4. They are non-conductors of heat and electricity. Exceptions: C and some others are conductors.

5. They are usually malleable and ductile.	5. They are deficient in malleability and ductility.
6. They form alloys, or "chemical mixtures," with one another, similar to other solutions. *Exceptions*: Some, as Pb and Zn, will not alloy with one another.	6. They often form liquid solutions, similar to alloys in metals.
7. Metals are electro-positive elements, and unite with O and H to form bases. *Exceptions*· Some of the less electro-positive metals, with a large quantity of O, form acids, as Cr, As, etc. Numbers 2, 6, and 7 are the most characteristic and important properties.	7. Non-metals are electro-negative, and with H, or with H and O, form acids.

Examine brass, bronze, bell-metal, pewter, German silver, solder, type-metal.

223. Alloys. — An alloy is not usually a definite chemical compound, but rather a mixture of two or more metals which are melted together. One metal may be said to dissolve in the other, as sugar dissolves in water. The alloy has, however, different properties from those of its elements. For example, plumber's solder melts at a lower temperature than either Pb or Sn, of which it is composed. Some metals can alloy in any proportions. Solder may have two parts of Sn to one of Pb, two of Pb to one of Sn, or equal parts of each, or the two elements may alloy in other proportions. Not all metals can be thus fused together indefinitely; *e.g.*, Zn and Pb. Nickel and silver coins are alloyed with Cu, gold coins with Cu and Ag.

Gun-metal, bell-metal, and speculum-metal are each alloys of Cu and Sn. Speculum-metal, used for reflectors in telescopes, has relatively more

Sn than either of the others; gun-metal has the least. An alloy of Sb and Pb is employed for type-metal as it expands at the instant of solidification. Pewter is composed of Sn and Pb; brass, of Cu and Zn; German silver, of brass and Ni; bronze, of Cu, Sn, and Zn; aluminium bronze, of Cu and Al.

224. Low Fusibility is a feature of many alloys. Wood's metal, composed of Pb eight parts, Bi fifteen, Sn four, Cd three, melts at just above 60°, or far below the boiling-point of water. By varying the proportions, different fusing-points are obtained. This principle is applied in automatic fire alarms, and in safety plugs for boilers and fire extinguishers. Water pipes extend along the ceiling of a building and are fitted with plugs of some fusible alloy, at short distances apart. When, in case of fire, the heat becomes sufficiently intense, these plugs melt and the water flows out.

225. Amalgams. — An amalgam is an alloy of Hg and another metal. Mirrors are "silvered" with an amalgam of Sn. Tin-foil is spread on a smooth surface and covered with Hg, and the glass is pressed thereon.

Various amalgams are employed for filling teeth, a common one being composed of Hg, Ag, and Sn. Au or Ag, with Hg, forms an amalgam used for plating. Articles of gold and silver should never be brought in contact with Hg. If a thin amalgam cover the surface of a gold ring or coin, Hg can be removed with HNO_3, as Au is not attacked by it. Would this acid do in case of silver amalgam? Heat will also quickly cause Hg to evaporate from Au.

CHAPTER XLIV.

SODIUM AND ITS COMPOUNDS.

Examine $NaCl$, Na_2SO_4, Na_2CO_3, Na, $NaOH$, $HNaCO_3$, $NaNO_3$.

226. Order of Derivation. — Though K is more metallic, or electro-positive, than Na, the compounds of Na are more important, and will be considered first. The only two compounds of Na which occur extensively in nature are $NaCl$ and $NaNO_3$. Almost all others are obtained from NaCl, as shown by this table, which should be memorized and frequently recalled.

$$NaCl \begin{cases} Na_2SO_4 \begin{cases} Na_2CO_3 \begin{cases} Na \\ NaOH \\ HNaCO_3 \end{cases} \end{cases} \end{cases}$$
$$NaNO_3$$

From what is Na_2SO_4 prepared, as shown by the table? Na_2CO_3? Na?

227. Occurrence and Preparation of NaCl. — NaCl occurs in sea water, of which it constitutes about three per cent, in salt lakes, whose waters sometimes hold thirty per cent, or are nearly saturated, and, as rock salt, in large masses underground. Poland has a salt area of 10,000 square miles, in some parts of which the pure transparent rock salt is a quarter of a mile thick. In Spain there is a mountain of salt five hundred feet high and three miles in circumference. France obtains much salt from sea water. At high tide it flows into shallow basins, from which the sun evaporates the water, leaving NaCl to crystallize. In Norway it is separated by freezing water, and in Poland it is mined like coal. In New York and Michigan it is obtained by evaporating the brine of salt wells, either by air and the sun's heat, or by fire. Slow evaporation gives large crystals; rapid, small ones.

228. Uses. — The main uses are for domestic purposes and for making the Na and Cl compounds. In the United States the consumption amounts to more than forty pounds per year for every person.

229. Sodium Sulphate. — What acid and what base are represented by Na_2SO_4? Which is the stronger acid, HCl or H_2SO_4? Would the latter be apt to act on $NaCl$? Why?

230. Manufacture. — This comprises two stages shown by the following reactions, in which the first needs moderate heat only; the last, much greater.
   (1) $2 NaCl + H_2SO_4 = HNaSO_4 + NaCl + HCl$.
   (2) $NaCl + HNaSO_4 = Na_2SO_4 + HCl$.

The operation is carried on in large furnaces. The gaseous HCl is passed into towers containing falling water in a fine spray, for which it has great affinity. The solution is drawn off at the base of the tower. Thus all commercial HCl is made as a by-product in manufacturing Na_2SO_4.

When crystalline, sodium sulphate has ten molecules of water of crystallization (Na_2SO_4, $10 H_2O$); it is then known as Glauber's salt. This salt readily effloresces; *i.e.* loses its water of crystallization, and is reduced to a powder. Compute the percentage of water.

231. Uses. — The leading use of Na_2SO_4 is to make Na_2CO_3; it is also used to some extent in medicine, and in glass manufacture.

232. Sodium Carbonate. — Note the base and the acid which this salt represents. Test a solution of the salt with red and blue litmus, and notice the alkaline reaction. Do you see any reason for this reaction in the strong base and the weak acid represented by the salt?

233. Manufacture. — Na_2CO_3 is not made by the union of an acid and a base, nor is H_2CO_3 strong enough to act on many salts. The process must be indirect. This consists in reducing Na_2SO_4 to Na_2S, by taking away the O with C, charcoal, and then changing Na_2S to Na_2CO_3 by $CaCO_3$, limestone. The three substances, Na_2SO_4, C, $CaCO_3$, are mixed together and strongly heated. The reactions should be carefully studied, as the process is one of much importance.

(1) $Na_2SO_4 + 4 C = Na_2S + 4 CO$.
(2) $Na_2S + CaCO_3 = CaS + Na_2CO_3$.

Observe that C is the reducing agent. The gas CO escapes. The solid products Na_2CO_3 and CaS form black ash, the former being very soluble, the latter only sparingly soluble in water. Na_2CO_3 is dissolved out by water, and the water is evaporated. This gives commercial soda. CaS, the waste compound in the process, contains the S originally in the H_2SO_4 used. This can be partially separated and again made into acid. Describe the manufacture of Na_2CO_3 in full, starting with NaCl. This is called the Le Blanc process, but is not the only one now employed to produce this important article.

234. Occurrence. — Sodium carbonate is found native in small quantities. It forms the chief surface deposit of the "alkali belt" in western United States, where it often forms incrustations from an inch to a foot in thickness.

SODIUM AND ITS COMPOUNDS. 141

It was formerly obtained from sea-weeds, by leaching their ashes, as, by a like process, K_2CO_3 was obtained from land plants.

235. Uses. — Na_2CO_3 forms the basis of many alkalies, as H_2SO_4 does of acids. Of all chemical compounds it is one of the most important, and its manufacture constitutes one of the greatest chemical industries. Its economical manufacture largely depends on the demand for HCl, which is always formed as a by-product. As but little HCl is used in this country, Na_2CO_3 is mostly manufactured in Europe. The chief uses are for glass (page 132) and alkalies.

236. Sodium. — Na must always be kept under naphtha, or some other liquid compound containing no O, since it oxidizes at once on exposure to the air. For this reason it never occurs in a free state.

237. Preparation. — By depriving Na_2CO_3 of C and O, metallic sodium is formed. As usual, heated charcoal is the reducing agent. The end of the retort, which holds the mixture, dips under naphtha.

$Na_2CO_3 + 2\,C = 2\,Na + 3\,CO$. The process is a difficult one, and Na brings five dollars per pound, though in its compounds it is a third as common as Fe. K is as abundant as Na, but more difficult of separation, and is worth three dollars per ounce. Notice the position of K and Na at the positive end of the elements, page 43.

238. Uses. — Na is used to reduce Al, Ca, Mg, Si, which are the most difficult elements to separate from their compounds. It acts in these cases as a reducing agent.

239. Sodium Hydrate.

Review Experiment 62, page 69.

Experiment 118. — Put into a t.t 10^{cc} H_2O and 2 or 3^g NaOH. Note its easy solubility. Test with litmus. Will it neutralize any acids? See page 53.

240. Preparation. — Sodium hydrate, caustic soda, or soda by lime, is made by treating a solution of Na_2CO_3 with milk of lime (page 69). $CaCO_3$ is precipitated and allowed to settle, the solution is poured off, and NaOH is obtained by evaporating the water and running the residue into molds.

241. Use. — NaOH is a powerful caustic, but its chief use is in making hard soap. See page 187.

242. Hydrogen Sodium Carbonate. — Hydrogen sodium carbonate, bicarbonate of sodium, acid sodium carbonate, cooking-soda, etc., $HNaCO_3$, is prepared by passing CO_2 into a solution of Na_2CO_3. $Na_2CO_3 + H_2O + CO_2 = 2\,HNaCO_3$. Test a solution of it with litmus. Account for the result. Its use in bread-making depends on the ease with which CO_2 is liberated. Even a weak acid, as the lactic acid of sour milk, sets this free, and thus causes the dough to rise.

243. Sodium Nitrate. — Sodium nitrate occurs in Chili and Peru. It is the main source of HNO_3.

Review Experiments 46 and 52. From $NaNO_3$ is also made KNO_3 ($NaNO_3 + KCl = NaCl + KNO_3$), one of the ingredients of gunpowder. By reason of its deliquescence $NaNO_3$ is not suitable for making gunpowder, though it is sometimes used for blasting-powder. The action of the latter is slower than that made from KNO_3. $NaNO_3$ is cheaper and more abundant than KNO_3; this is true of most Na compounds in comparison with those of K.

CHAPTER XLV.

POTASSIUM AND AMMONIUM.

POTASSIUM AND ITS COMPOUNDS.

Examine K, KCl, K_2SO_4, K_2CO_3, KOH, $HKCO_3$, $KClO_3$, KCN.

244. Occurrence and Preparation. — Potassium occurs only in combination, chiefly as silicates, in such minerals as feldspar and mica. By their disintegration it forms a part of soils from which such portions as are soluble are taken up by plants. The ashes of land-plants are leached in pots to dissolve K_2CO_3; hence it is called *potash*. Sea-plants likewise give rise to Na_2CO_3. Wood ashes originally formed the main source of K_2CO_3. From plants this substance is taken into the animal system, and makes a portion of its tissue. Sheep excrete it in sweat, which is then absorbed by their wool. Large quantities are now obtained by washing wool and evaporating the water.

K_2CO_3 and other compounds of K are mainly derived from KCl, beds of which exist in Germany.

In the following list each K compound is prepared like the same Na compound, and the uses of each of the former are similar to those of the latter. K compounds are made in much smaller quantities than those of Na, as KCl is far less common than NaCl.

$$\begin{matrix} KCl \\ KNO_3 \end{matrix} \Big\{ K_2SO_4 \Big\{ K_2CO_3 \Big\{ \begin{matrix} K \\ KOH \\ HKCO_3 \end{matrix}$$

Examine specimens of each, side by side with like Na compounds. Describe in full their preparation, giving the reactions. Also, perform the experiments given under Na, substituting K therefor. From KOH are made KClO₃ and KCN.

$$\text{KOH} \begin{cases} \text{KClO}_3 \\ \text{KCN} \end{cases}$$

245. Potassium Chlorate. — $KClO_3$ is made by passing Cl into a hot concentrated solution of KOH.

$$6\ KOH + 6\ Cl = KClO_3 + 5\ KCl + 3\ H_2O.$$

Its uses are in making O, and as an oxidizing agent.

246. Potassium Cyanide, KCN, is a salt from HCN — hydrocyanic or prussic acid. Each is about equally poisonous, and more so than any other known substance. A drop of pure HCN on the tongue will produce death quickly by absorption into the system. In examining these compounds take care not to handle them or to inhale the fumes. KCN is used as a solvent for metals in electro-plating, and is the source of many cyanides, *i.e.* compounds of CN and a metal. KCN is employed to kill insects for cabinet specimens. In a wide-mouthed bottle is placed a little KCN, which is covered with cotton, and over this a perforated paper. The bottle is inverted over the insect, and the fumes destroy life without injuring the delicate parts. HCN is made from KCN and H_2SO_4.

247. Gunpowder. — Gunpowder is a *mixture* of KNO_3, C, and S. Heat or concussion causes a chemical change, and transforms the solids into gases. These gases at the moment of explosion occupy 1500 or more times the volume of the solids. Hence the great rending power of powder. If not confined, powder burns quietly but quickly. The appended reaction is a part of what takes place, but it by no means represents all the chemical changes.

$$2\ KNO_3 + S + 3\ C = K_2S + 2\ N + 3\ CO_2.$$

From this equation compute the percentage, by weight, of each substance used to make gunpowder economically.

Thoroughly burned charcoal, distilled sulphur, and the purest nitre are powdered and mixed in a revolving drum, made into a paste with water, put under great pressure between sheets of gun metal, granulated, sifted, to separate the coarse and fine grains, and glazed by revolving in a barrel which sometimes contains a little powdered graphite.

Experiment 119. — Pulverize and mix intimately 4ᵍ KNO_3, ½ᵍ S, ½ᵍ charcoal. Pile the mixture on a brick, and apply a lighted match. The adhering product can be removed by soaking in water.

AMMONIUM COMPOUNDS.

248. Read the chapter on NH_3. Also, review the experiments on bases. Examine NH_4Cl, NH_4NO_3, $(NH_4)_2SO_4$, $(NH_4)_2CO_3$.

Ammonium, NH_4, is too unstable to exist alone, but it forms salts similar to those of K and Na. NH_3 dissolved in water forms NH_4OH.

The food of plants, as well as that of animals, must contain N. It has not yet been shown that they can make use of that contained in the air, but they do absorb its compounds from the soil. All fertilizers and manures contain a soluble compound of NH_4. All NH_4 compounds are now obtained either from coal, in making illuminating-gas, or from bones, by distillation.

Suppose the product obtained from the gas-house to be NH_4OH, how would NH_4Cl be made? $(NH_4)_2SO_4$? NH_4NO_3? Write the reactions. $(NH_4)_2CO_3$ is made by heating NH_4Cl with $CaCO_3$. Give the reaction.

CHAPTER XLVI.

CALCIUM COMPOUNDS.

Examine $CaCO_3$ — marble, limestone, chalk, not crayon, — $CaSO_4$ — gypsum or selenite — $CaCl_2$, CaO.

249. Occurrence. — The above are the chief compounds of Ca. The element itself is not found uncombined, is very difficult to reduce (page 141), is a yellow metal, and has no use. Its most abundant compound is $CaCO_3$. Shells of oysters, clams, snails, etc., are mainly $CaCO_3$, and coral reefs, sometimes extending thousands of miles in the ocean, are the same. $CaCO_3$ dissolves in water holding CO_2, and thence these marine animals obtain it and therefrom secrete their bony framework. All mountains were first laid down on the sea bottom layer by layer, and afterwards lifted up by pressure. Rocks and mountains of $CaCO_3$ were formed by marine animals, and all large masses of $CaCO_3$ are thought to have been at one time the framework of animals. Marble is crystallized, transformed limestone. The process, called metamorphism, took place in the depths of the earth, where the heat is greater than at the surface.

250. Lime. — If $CaCO_3$ be roasted with C, CO_2 escapes and CaO is left. $CaCO_3 - CO_2 = ?$ This is called burning lime, and is a large industry in limestone countries. CaO is unslaked lime, quicklime or calcium oxide. It may be slaked either by exposure to the air,

air-slaking, when it gradually takes up H_2O and CO_2; or by mixing with H_2O, water-slaking. $CaO + H_2O = Ca(OH)_2$. Great heat is generated in the latter case, though not so much as in the formation of KOH and NaOH. Like them, $Ca(OH)_2$ dissolves in water, forming lime-water. Milk of lime, cream of lime, etc., consist of particles of $Ca(OH)_2$ suspended in H_2O.

251. Uses of Lime. — CaO is infusible at the highest temperatures. If it be introduced into the oxy-hydrogen blow-pipe (page 28), a brilliant light, second only to the electric, is produced. Mortar is made by mixing CaO, H_2O, and SiO_2. It hardens by evaporating the extra H_2O, absorbing CO_2 from the air, and uniting with SiO_2 to form calcium silicate. It often continues to absorb CO_2 for hundreds or thousands of years before being saturated, as is found in the Egyptian pyramids. Hence the tenacity of old mortar. Hydraulic mortar contains silicates of Al and Ca, and is not affected by water. What are the uses of mortar? Being the important constituent of mortar and plaster, lime is the most useful of the bases.

252. Hard Water. — Review Experiment 76. The solubility of $CaCO_3$ in water that contains CO_2 leads to important results. Much dissolves in the waters of all limestone countries; and the water, though perfectly transparent, is hard; *i.e.* soap has little action on it. See page 187. Such water may be softened by boiling, a deposit of $CaCO_3$ being formed as a crust on the kettle. Such water is called water of temporary hardness. $MgCO_3$ produces a similar effect, and water containing it is softened in the same way. Permanently hard waters contain the sulphates of Ca and Mg, which cannot be removed by boiling, but may be by adding $(NH_4)_2CO_3$.

253. The Formation of Caves in limestone rocks is due also to the solubility of $CaCO_3$. Water collects on the mountains and trickles down through crevices, dissolving, if it contains CO_2, some of the $CaCO_3$, and thus making a wider opening, and forcing its way along fissures and lines of least resistance into the interior of the earth, or out at the base of the mountain. Its channel widens as it dissolves the rock, and the stream enlarges until in the course of ages an immense cavern may be formed, with labyrinths extending for miles, from the entrance of which a river often issues. In the long ages which elapsed during the slow formation of Mammoth Cave its denizens lost many of the characters of their ancestors, and eyeless fish and also eyeless insects now abound there.

254. Reverse Action. — Drops of water on the roofs of these caverns lose their CO_2 and deposit $CaCO_3$. Thus long, pendant masses of limestone, called stalactites, are slowly formed on the roofs like icicles. From these, water charged with $CaCO_3$ drops to the bottom, loses CO_2 and deposits $CaCO_3$, which forms an upward-growing mass, called stalagmite. In time it may meet the stalactite and form a pillar. Notice that the same action which formed the cave is filling it up; *i.e.* the solubility of $CaCO_3$ in water charged with CO_2.

255. Famous Marbles. — The marble from Carrara, Italy, is most esteemed on account of a pinkish tint given by a trace of oxide of iron. The best of Grecian marble was from Paros, one of the Cyclades. The isles of the Mediterranean are of limestone, or of volcanic, origin, often of both.

256. Calcium Sulphate occurs in two forms, (1) with water of crystallization — gypsum, $CaSO_4 + 2\,H_2O$, — (2) without it — anhydrite, $CaSO_4$. The former, on being strongly heated, gives up its water, and is reduced to a powder — plaster of Paris. This, on being mixed with water, again takes up $2\,H_2O$, and hardens, or sets, without crystallizing. If once more heated to expel water, it will not again absorb it. When plaster of Paris sets, it expands slightly, and on this account is admirable for taking casts.

257. Uses. — Gypsum finds use as a fertilizer and as an adulterant in coloring-materials, etc.

$CaSO_4$ is employed in making casts, molds, statuettes, wall-plaster, crayons, etc.

How can $CaCl_2$ be made? What is its use? See page 27. What else is used for a similar purpose?

Symbolize and name the acid represented by $Ca(ClO)_2$, and name this salt (page 107). It is one of the constituents of bleaching-powder, the symbol of which, though still under discussion, may be considered $Ca(ClO)_2 + CaCl_2$. This is made by passing Cl over $Ca(OH)_2$, $2\,Ca(OH)_2 + 4\,Cl = Ca(ClO)_2 + CaCl_2 + 2\,H_2O$.

CHAPTER XLVII.

MAGNESIUM, ALUMINIUM, AND ZINC.

MAGNESIUM AND ITS COMPOUNDS.

Examine magnesite, dolomite, talc, serpentine, hornblende, meerschaum, magnesium ribbon, magnesia alba, Epsom salt.

258. Occurrence and Preparation. — Mg is very widely distributed, but does not occur uncombined. Its salts are found in rocks and soils, in sea water and in the water of some springs, to which they impart a brackish taste.

The most common minerals containing Mg are magnesite, $MgCO_3$, dolomite, $MgCO_3 + CaCO_3$, and talc, serpentine, hornblende, and meerschaum. The last four are silicates, and often are unctious to the touch. What proportion of the earth's crust is composed of Mg? See page 173.

259. Metallic Mg is prepared by fusing $MgCl_2$ with Na. Why is the process expensive? Write the reaction.

Experiment 120. — With forceps hold a short strip of Mg ribbon in a flame. Note the brilliancy of the light, and give the reaction. Examine and name the product.

Photographs of the interior of caverns, where sunlight does not penetrate, are taken by Mg light. Gun-cotton sprinkled with powdered Mg has recently been employed for that purpose. Mg tarnishes slightly in moist air.

Compounds of Mg. — MgO, magnesia, like CaO, is very infusible, and is used for crucibles. Magnesia alba, a variable mixture of $MgCO_3$ and $Mg(OH)_2$, is employed in medicine, as is also Epsom salt, $MgSO_4 + 7 H_2O$.

ALUMINIUM AND ITS COMPOUNDS.

Examine aluminium, aluminium bronze, corundum, emery, feldspar, argillite, clay. Note especially the color, luster, specific gravity and flexibility of Al.

What elements are more common in the earth than Al? What metals? See page 173. Compare the abundance of Al with that of Fe.

260. Compounds of Al. — Al occurs only in combination with other elements. Feldspar, mica, slate, and clay are silicates of it. It occurs in all rocks except $CaCO_3$ and SiO_2, and in nearly 200 minerals. Though found in all soils, its compounds are not taken up by plants, except by a few cryptogams. Corundum, Al_2O_3, is the richest of its ores. Compute its per cent of Al. Compounds of Al are very infusible and difficult of reduction.

261. Reduction. — Like most other metals not easily reducible by C or H, it was originally obtained by electrolysis, but more recently from its chloride, by the reducing action of strongly heated K or Na. $Al_2Cl_6 + 6 Na = 6 NaCl + 2 Al$.

What is the chief use of Na? See page 141. As it takes three pounds of Na to make one pound of Al, the cost of the latter has been fifteen dollars or more per pound. Its use has thus been restricted to light apparatus and aluminium bronze, an alloy of Cu 90, Al 10, which is not unlike gold in appearance.

Al_2O_3 has lately been reduced by C. Higher temperatures than have heretofore been known are obtained by means of the electric arc and large dynamo machines. A furnace made of graphite, because fire-clay melts like wax at such a high temperature, is filled with Al_2O_3 — corundum, — C, and Cu. In the midst of this are embedded large carbon terminals, connected with dynamos. The reduction takes several hours.

The following reaction takes place: $Al_2O_3 + 3\,C = 2\,Al + 3\,CO$. Cu is also added, and an alloy of Al and Cu is thus formed. This alloy is not easily separable into its elements. Explain the action of the C. CO escapes through perforations in the top of the furnace, burning there to CO_2. Only alloys of Al have yet been obtained by this process. This method has not been employed before, simply because the highest temperatures of combustion, 2000° or 2500°, would not effect a reduction. In the same way Si, B, K, Na, Ca, Mg, Cr, have recently been reduced from their oxides; but a process has yet to be found for separating them easily from their alloys.

262. Properties and Uses. — Al is a silvery white metal, lighter than glass, and only one-third the weight of iron. It does not readily rust or oxidize, it fuses at 1000° (compare with Fe), is unaffected by acids, except by HCl and, slightly, by H_2SO_4, is a good conductor of electricity, can be cast and hammered, and alloys with most metals, forming thus many valuable compounds. Every clay-bank is a mine of this metal, which has so many of the useful properties of metals and has so few defects that, if it could be obtained in sufficient quantities, it might, for many purposes, take the place of iron, steel, tin, and other metals. From its properties state any advantages which it would have over iron in ocean vessels, railroads, and bridges. Why is it better than Sn or Cu for culinary utensils?

An alloy of Al, Cu, and Si is used for telephone wires in Europe, and the Bennett-Mackay cable is of the same material. Washington monument, the tallest shaft in the world, is capped with a pyramid of Al, ten inches high.

For the uses of alumina, Al_2O_3, and its silicates, see page 133.

ZINC AND ITS COMPOUNDS.

Examine zincite, sphalerite, Smithsonite, sheet zinc, galvanized iron, granulated zinc, zinc dust.

263. Compounds. — The compounds of zinc are abundant. Its chief ores are zincite, ZnO, sphalerite or blende, ZnS, Smithsonite, $ZnCO_3$. For their reduction these ores are first roasted, *i.e.* heated in presence of air. With ZnS this reaction takes place: $ZnS + 3\,O = ZnO + SO_2$. The oxide is reduced with C, and then Zn is distilled. State the reaction. Zinc is sublimed — in the form of zinc dust — like flowers of S. Granulated Zn is made by pouring a stream of the molten metal into water.

Experiment 121. — Burn a strip of Zn foil, and note the color of the flame and of the product. State the reaction. The red color of zincite is supposed to be imparted by Mn present in the compound.

264. Uses. — Name any use of Zn in the chemical laboratory. It is employed for coating wire and sheet iron — galvanized iron. This is done by plunging the wire or the sheets of iron into melted Zn. Describe the use of Zn as an alloy. See page 136.

ZnO forms the basis of a white paint called zinc white. White vitriol, $ZnSO_4 + 7\,H_2O$, is employed in medicine. Name two other vitriols.

CHAPTER XLVIII.

IRON AND ITS COMPOUNDS.

Examine magnetite, hematite, limonite, siderite, pig-iron, wrought-iron, steel.

265. Ores and Irons. — As Fe occurs native only in meteorites and in small quantities of terrestrial origin, it is obtained from its ores. There are four of these ores — magnetite (Fe_3O_4), hematite (Fe_2O_3), limonite ($2\ Fe_2O_3 + 3\ H_2O$), and siderite ($FeCO_3$). Which is richest in Fe? Compute the proportion. $FeCO_3$ occurs mostly in Europe. The reduction of these ores, as well as of other metallic oxides, consists in removing O by C at a high temperature. As ordinarily classified there are three kinds of iron, — pig- or cast-iron, steel, and wrought-iron.

Study this table, noting the purity, the fusing-point, and the per cent of C in each case.

	Per Cent Fe (general).	Fusibility.	Per Cent C.
Pig	90	1200°	2–6
Steel	99	1400°	0.5–2
Wrought	99.7	1500°	Fraction.

Pure iron melts at about 1800°. Pig-iron is obtained from the ore by smelting, and from this are made steel and wrought-iron.

266. Pig-Iron. — The ore is reduced in a blast furnace (Fig. 47), in some cases eighty or one hundred feet high, and having a capacity of about 12,000 cubic feet. The reducing agent is either charcoal, anthracite coal, or coke, bituminous coal being too impure. Charcoal is the best agent, and is used in preparing Swedish iron; but it is too expensive for general use.

Fig. 47.

Blast furnace. F, entrance of tuyeres, or blast-pipes. E, F, hottest part. C, conductor for gases, which are subsequently used to heat the air going into the tuyeres. G, upper portion, slag, lower portion, melted iron.

Were ores absolutely pure, only C would be needed to reduce them. Complete: $Fe_3O_4 + 4\,C = ?$ $Fe_3O_4 + 2\,C = ?$

Much earthy material — gangue — containing silica and silicates is always found with iron ores. These are infusible, and something must be added to render them fusible. CaO forms with SiO_2 just the flux needed. See page 132. $CaO + SiO_2 = ?$ Which of these is the basic, and which the acidic compound? CaO results from heating $CaCO_3$; hence the latter is employed instead of the former. In what case would SiO_2 be used as the flux?

Into the blast furnace are put, in alternate layers, the fuel, the flux, and the ore. The fire, once kindled, is kept burning for months or years. Hot air is driven in through the tuyeres (*tweers*). O unites with C of the fuel, forming CO_2 and CO. The C also reduces the ore. $Fe_2O_3 + 3\,C = ?$ CO accomplishes the same thing. $3\,CO + Fe_2O_3 = ?$ The intense heat fuses CaO and SiO_2 to a silicate which, with other impurities, forms a slag; this, rising to the surface of the molten mass, is drawn off. The iron is melted, falls in drops to the bottom, and is drawn off into sand molds. See Figure 47. This is pig-iron. It contains as impurities, C, Si, S, P, Mn, etc. If too much S or P is present in an ore, it is worthless. This is why the abundant mineral FeS_2 cannot be used as a source of iron. From the top of the furnace N, CO, CO_2, H_2O, etc., escape. These gases are used to heat the air which is forced through the tuyeres, and to make steam in boilers.

267. Steel. — The manufacture of steel and wrought-iron consists in removing most of the impurities from pig-iron. It will be seen that the most common compounds of C, S, Si, and P, are their oxides, and these are for the most part gases. Hence these elements are removed by oxidation.

Bessemer steel is prepared by melting pig-iron and blowing hot air through it. A converter (Fig. 48) lined with siliceous sand, and holding several tons, is partially filled with the molten metal; blasts of hot air are driven into it, and the C and other impurities, together with a little of the Fe, are oxidized. The exact moment when the process has gone far enough, and most of the impurities have been removed, is indicated by the appearance of the escaping flame. It usually takes from five to ten minutes. The blast is then stopped, and the metal has about the composition of wrought-iron; it contains some uncombined O. A white pig-iron (spiegeleisen), which contains a known quantity of C and of Mn, is at once added. Mn removes part of the extra O, and, though it remains, does not injure the metal. The C is "dissolved" by the Fe, which is then run into molds (ingots). This process, the Bessemer, invented in 1856, has revolutionized steel manufacture. No less than ten tons of iron have been converted into steel, in five minutes, in a single converter.

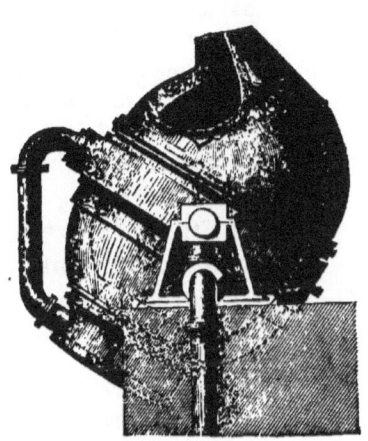

Fig. 48.

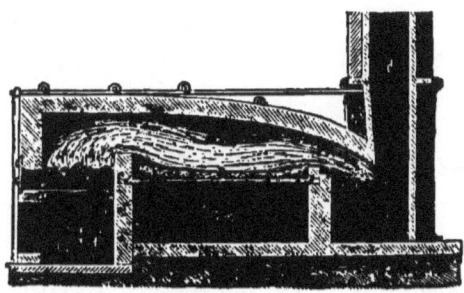

Fig. 49.

268. Wrought-Iron. — The chemical principle involved in making wrought-iron is the same as that in making steel, but the process is different. Impurities are burned out from pig-iron in an open reverberatory furnace, by constantly stirring the metal in contact with air. This is called puddling. A reverberatory furnace is one in which the fuel is in one compartment, and the heat is reflected downward into another, that holds the substance to be acted upon (Fig. 49).

Steel may also be made by carburizing wrought-iron. Iron and charcoal are packed together and heated for days, without melting, when it is found that, in some unknown way, solid C has penetrated solid Fe. The finer kinds of steel are made in this way, but they are very expensive.

Wrought-iron may also be made directly from the ore in an open hearth furnace, with charcoal. This was the original mode.

269. Properties. — The varying properties of pig-iron, steel, and wrought-iron are due in part to the proportion of C and of other elements present, either as mixtures or as compounds, and in part to other causes not well understood. Wrought-iron is fibrous, as though composed of fine wires, and hence is ductile, malleable, tough, and soft, and cannot be hardened or tempered, but it is easily welded. Pig-iron is crystalline, and so is not ductile or malleable; it is hard and brittle, and cannot be welded. On account of its low melting-point it is generally employed for castings. Steel is crystalline in structure, and when suddenly cooled from red heat by plunging into cold water, becomes hard and brittle. The tempering can be varied by afterwards heating to any required degree, indicated by the color of the oxide formed on the exterior. The higher temperatures give the softer steel.

270. Salts of Iron. — Examine $FeSO_4$, FeS, FeS_2.

Fe has a valence of 2 or 4. This gives rise to two kinds of salts, ferrous and ferric, as in $FeCl_2$ and Fe_2Cl_6. The valence of Fe in ferric salts is 4. See page 40. Ferrous sulphate is $FeSO_4$; ferric sulphate, $Fe_2(SO_4)_3$. Write the symbols for ferrous and ferric hydrate; for the oxides; for the nitrates. Write the graphic symbols for each.

IRON AND ITS COMPOUNDS. 159

271. Colors. — The characteristic color of ferrous salts is green, as in $FeSO_4$. These salts give the green color to the chlorophyll in leaves and grass, and bottle glass owes its green color to ferrous silicate. Ferric salts are a brownish red, as shown in hematite and limonite, and in some bottles. Red sandstone, and most soils and earths, are illustrations of this coloring action. The blood of vertebrates owes its color to ferric salts. Bricks are made from a greenish blue clay in which iron exists in the ferrous state. On being heated, ferrous salts are oxidized to ferric, and their color is changed to red. Iron rust is hydrated ferric oxide, Fe_2O_3 and $Fe_2(OH)_6$.

272. Change of Valence.

Experiment 122.—Dissolve 2^g of iron filings in diluted HCl. Filter or pour off the clear liquid, divide it into two parts, and add NH_4OH to one part till a ppt. occurs. Notice the greenish color of $Fe(OH)_2$. Oxidize the other part by adding a few drops of HNO_3 and boiling a minute. Now add NH_4OH, and observe the reddish color of the ppt., $Fe_2(OH)_6$.

Solutions of ferrous salts will gradually change to ferric, if allowed to stand, thus showing the greater stability of the latter. In changing from $FeCl_2$ to Fe_2Cl_6 oxidation does not consist in adding O, but in increasing the negative element or radical. This is possible only by changing the valence of Fe from 2 to 4. Hence oxidation, in its larger sense, means increasing the valence of the positive element. To oxidize $FeSO_4$ is to make it $Fe_2(SO_4)_3$, changing the valence of Fe as before. Reduction or deoxidation diminishes the valence of the positive element. Illustrate this by the same iron salts. Illustrate it by PbO and PbO_2; $AuCl$ and $AuCl_3$; Sb_2S_3 and Sb_2S_5. In this sense define an oxidizing agent. A reducing agent.

273. Ferrous Sulphate.

Experiment 123. — Dissolve a few iron filings in dilute H_2SO_4, and slowly evaporate for a few minutes. Write the equation.

Ferrous sulphate, green vitriol, or copperas, $FeSO_4 + 7 H_2O$, is the source of what acid? See page 66. It is also one of the ingredients in many writing inks. On being heated, or exposed to the air, it loses its water of crystallization and becomes a white powder. It is prepared as above, or by oxidizing moistened FeS_2 by exposure to the air.

Ferrous sulphide, protosulphide of iron, FeS, is how prepared? See Experiment 6. State its use. See Experiment 108. It also occurs native.

Ferric sulphide, pyrite, FeS_2, occurs native in large quantities. What is its use? See page 65.

CHAPTER XLIX.

LEAD AND TIN.

LEAD.

Examine galena, lead protoxide and dioxide, red-lead, lead carbonate, acetate, and nitrate. Note especially the colors of the oxides, the cubical crystallization and cleavage of galena, the specific gravity of the compounds, the softness of Pb, and the tarnish, Pb_2O, which covers it, if long exposed.

274. Distribution of Pb. — Pb is widely distributed, occurring as PbS and $PbCO_3$. PbS, galenite or galena, is its main source. By heating it in air, SO_2 is formed, and Pb liberated and drawn off.

Pb is but little acted on by cold H_2SO_4, unless concentrated. Describe its use in making that acid. See page 65. To show that a little Pb has been dissolved, as $PbSO_4$, in the manufacture of that acid, perform this experiment.

Experiment 124. — To 5^{cc} of water in a clean t.t. add the same volume of H_2SO_4, not C.P.; shake, and notice any fine powder suspended. $PbSO_4$, being insoluble in water, is precipitated. What is the test for Pb? See Experiment 109.

275. Poisonous Properties. — Pb is very flexible and soft, and is much used for water pipes. In moist air it is soon coated with suboxide, Pb_2O, as may be seen by exposing a fresh surface. Some portion of this is liable to dissolve in water, and, as all soluble salts of Pb are poisonous, water that has stood in pipes should not be used for drinking. Lead is employed as an alloy of tin for covering sheet-iron in "terne plate." This plate is rarely used except for roofing. The "bright plate," used for tin cans and other purposes, scarcely ever contains any lead except the small portion in solder. In soldering, $ZnCl_2$ is employed for a flux. Sn, Pb, and Zn are somewhat soluble in vegetable acids. If citric acid be present, as it usually is, citrates of these metals are formed, and all of them are poisonous. The action is far more rapid after opening the can, since oxidation is hastened. Hence the contents should be taken out directly after opening.

Lead poisons seem to have an affinity for the tissues of the body, and accumulate little by little. Painter's colic results from lead poisoning. Epsom salt, or other soluble sulphate, is an antidote, since with Pb it makes insoluble $PbSO_4$.

276. Some Lead Compounds. — Lead salts form the basis of many paints. White paint is a mixture of $PbCO_3$ and $Pb(OH)_2$ suspended in linseed oil. It is often adulterated with $BaSO_4$, ZnO, $CaCO_3$. Other lead compounds are used for colored paints. The two chief soluble salts are $Pb(NO_3)_2$ and lead acetate, $Pb(C_2H_3O_2)_2$.

Red-lead, Pb_3O_4, and, to some extent, litharge, PbO, are employed in glass manufacture. Name the kind of glass in which it is used, describe its manufacture, and write a symbol for lead silicate. What is the characteristic of lead glass? See page 132.

Experiment 125. — Put a small fragment of Pb on a piece of charcoal, and blow the oxidizing flame against it for some time with a mouth blow-pipe. Note the color of the coating on the coal. PbO has formed.

Experiment 126. — Dissolve a small piece of lead in dilute HNO_3. Pour off the solution into a t.t. and add HCl or other soluble chloride. $Pb(NO_3)_2 + 2\,HCl = ?$ What is the insoluble product?

Experiment 127. — Add to a solution of $Pb(C_2H_3O_2)_2$ some H_2SO_4. Give the reaction and the explanation.

TIN.

Examine cassiterite, tin foil, "terne plate," "bright plate."

277. Sn occurs as the mineral cassiterite, tin stone, SnO_2, and is found in only a few localities, as Banca, Malacca, and England. It does not readily tarnish, and is used to cover thin plates of copper and iron. Tin foil is generally an alloy of Pb and Sn.

Sn is sometimes a dyad, at others a tetrad. Write symbols for its two chlorides, stannous and stannic, also for its sulphides and oxides.

CHAPTER L.

COPPER, MERCURY, AND SILVER.

COPPER.

Examine native copper, chalcopyrite, malachite, azurite, copper acetate, copper nitrate, copper sulphate.

278. Occurrence. — Copper occurs both native and in many compounds, being diffused in rocks and, in minute quantities, in soils, waters, plants, and animals. Spain, Chili, and the United States are the chief Cu producing countries. The extensive mines of Michigan yield the native ore. The Calumet and Hecla mine alone produces 4,000,000 pounds per month. The most abundant compound of Cu is chalcopyrite, or copper pyrites, $CuFeS_2$. Malachite, which is green, and azurite, which is blue, are carbonates, the former being used for ornamental purposes.

Cu is, next to Ag, the best conductor of electricity and heat among the elements; it is very ductile, malleable, and tenacious.

Cu has two valences, 1 and 2. Symbolize and name its chlorides, iodides, sulphides, and oxides. Cupric compounds, as a rule, are more stable than cuprous.

279. Uses. — Thousands of tons of Cu find use in domestic utensils, ocean vessels, electric wires, batteries, and plating. Name the chief alloys of Cu and their uses. See page 136.

How may CuS be obtained? See Experiment 7. Cu_2O, cuprous oxide, is used to color glass red. $CuSO_4$ is employed in calico-printing, electric batteries, etc. It is called blue vitriol.

Paris green, used for killing potato-beetles, is composed chiefly of copper arsenite. Write the symbol for this compound. All soluble salts of Cu are poisonous; hence care should be taken not to bring any acid in contact with copper vessels of domestic use. With acetic acid, what would be formed?

MERCURY AND ITS COMPOUNDS.

Examine cinnabar, vermilion, mercury, red oxide, mercurous and mercuric chloride.

280. Cinnabar, HgS, is practically the only source of mercury — quicksilver. Austria, Spain, and California contain nearly all the mines. In these mines the metal also occurs native to a small extent. It is the only commonly occurring metal that is liquid at ordinary temperatures; it solidifies at about $-40°$. What other common liquid element? See page 12. Hg is reduced from the ore by Fe, Hg being distilled over and collected in water. Heat regularly expands the metal.

281. Uses. — For uses see Reduction of Ag and Au, pages 165 and 170; amalgams, page 137; laboratory work, page 68. It is also employed for thermometers and barometers, and as the source of the red pigment vermilion, which is artificial HgS.

Compare the vapor density and the atomic weight of Hg, and explain. See page 12. Hg is either a monad or a dyad. Symbolize its *ous* and *ic* oxides and chlorides. Which of the following are *ic* salts, and which are *ous*, and why? $HgNO_3$, $Hg(NO_3)_2$, HgCl, $HgCl_2$? Calomel, HgCl or Hg_2Cl_2, used in medicine, and corrosive sublimate, $HgCl_2$, are illustrations of the *ous* and *ic* salts. The former is insoluble, the latter soluble. All soluble compounds of Hg are virulent poisons, for which the antidote is the white of egg, albumen. With it they coagulate or form an insoluble mass.

SILVER AND ITS COMPOUNDS.

282. Occurrence and Reduction. — Silver is found uncombined, and combined, as Ag_2S, argenite, and AgCl, horn silver. It occurs usually with galena, PbS. It is abundant in the Western States, Mexico, and Peru. Silver is separated from galena by melting the two metals. As they slowly cool, Pb crystallizes, and is removed by a sieve, while Ag is left in the liquid mass. The principle is much like crystallizing NaCl from solution and leaving behind the salts of Mg, etc., in the mother liquor. When, by repeating the process, most of the Pb is eliminated, the rest is oxidized by heating in the air. $Pb + O = PbO$. Ag does not oxidize, and is left in the metallic state.

Another mode of reduction is to change the silver salt to its chloride, and then remove the Cl with Fe. Roasting with NaCl makes the first change, $2 NaCl + Ag_2S = Na_2S + 2 AgCl$, and with Fe the second, $2 AgCl + Fe = FeCl_2 + 2 Ag$. Ag is separated from the other products by adding Hg, with which it forms an amalgam. By distilling this, Hg passes over and Ag remains. This is the amalgamating process.

283. Salts of Silver are much employed in organic chemistry, and AgCl, AgBr, and $AgNO_3$ are used in photography. $AgNO_3$ is a soluble, colorless crystal, and is the basis of the silver salts. It blackens when in contact with organic matter. Stains on a photographer's hands are due to this substance, and the use of $AgNO_3$ in indelible inks depends on the same property. This may be due to a reduction of $AgNO_3$ to Ag_4O. Stains can be removed from the skin or from linen by a solution of KI, or of $CuCl_2$ followed by sodium hyposulphite. *Lunar caustic* is made by fusing $AgNO_3$ crystals, and is used for cauterizing (burning) the flesh. Much AgCN finds use in electroplating.

Experiment 128. — Put 5^{cc} $AgNO_3$ solution in each of three t.t. To the first add 3^{cc} HCl, to the second 3^{cc} NaCl solution, and to the third 3^{cc} KBr solution. Write the reaction for each case, and notice that the first two give the same ppt., as in fact any soluble chloride would. Filter the second and third, on separate filter papers, and expose half the residue to direct sunlight, observing the change of color by occasionally stirring. Solar rays reduce AgCl and AgBr, it is thought, to Ag_2Cl and Ag_2Br. Try to dissolve the other half in $Na_2S_2O_3$, sodium thiosulphate solution. This experiment illustrates the main facts of photography.

CHAPTER LI.

PHOTOGRAPHY.

284. Descriptive. — The silver halogens, AgCl, AgBr, AgI, are very sensitive to certain light rays. Red rays do not affect them; hence ruby glass is used in the "dark room."

Photography involves two processes. The negative of the picture is first taken upon a prepared glass plate, and the positive is then printed on prepared paper. The negative shows the lights and shades reversed, while the positive gives objects their true appearance.

Few photographers now make their own plates, these being prepared at large manufactories. The glass is there covered on one side with a white emulsion of gelatine and AgBr, making what are called gelatine-bromide plates. This is done in a room dimly lighted with ruby light. The plates are dried, packed in sealed boxes, and thus sent to photographers. The artist opens them in his dark room, similarly lighted, inserts the plates in holders, film side out, covers with a slide, adjusts to the camera, previously focused, and makes the exposure to light. The time of exposure varies with the kind of plate, the lens, and the light, from several seconds, minutes, or hours, to $\frac{1}{250}$ part of a second in some instantaneous work. In the dark room the plates are removed and can be at once developed, or kept for any time away from the light. No change appears in the plate until development, though the light has done its work.

To develop the plate, it is put into a solution of pyrogallic acid, the developer, and carbonate of sodium, the motive power in the process. Other developers are often used. The chemical action here is somewhat obscure, but those parts of the plates which were affected by the light are made visible, a part of the Ag_2Br being reduced to Ag by the affinity which sodium pyrogallate has for Br. $Ag_2Br = 2\,Ag + Br$. Br is dissolved and Ag is deposited. When the rather indistinct image begins to fade out, the plate is dipped for a minute into a solution of alum to harden the gelatine and prevent it from peeling off (frilling). It is finally soaked in a solution of sodium thiosulphate (hyposulphite or hypo), $Na_2S_2O_3$. This removes the AgBr that the light has failed to reduce. The process is called fixing, as the plate may thereafter be exposed to the light with impunity. It must be left in this bath till all the white part, best seen on the back of the plate, disappears. $2\,AgBr + 3\,Na_2S_2O_3 = Ag_2Na_4(S_2O_3) + 2\,NaBr$. Both products are dissolved. It is then thoroughly washed. Any dark objects become light in the negative, and *vice versa*. Why?

For the positive, the best linen paper is covered on one side with albumen, soaked in NaCl solution, dried, and the same side laid on a solution of $AgNO_3$. What reaction takes place? What is deposited on the paper, and what is dissolved? This sensitized paper, when dry, is placed over a negative, film to film, and exposed in a printing frame to direct sunlight till much darker than desired in the finished picture. What is dark in the negative will be light in the positive. Why? The reducing action of sunlight is similar to that in the negative. Explain it.

After printing, the picture is toned and fixed. Toning consists in giving it a rich color by replacing part of the Ag_2Cl with gold from a neutral solution of $AuCl_3$. $3\,Ag_2Cl + AuCl_3 = 6\,AgCl + Au$. Fixing removes the unaffected AgCl, as in the negative, the same substance being used. Describe the action. $2\,AgCl + 3\,Na_2S_2O_3 = Ag_2Na_4(S_2O_3)_3 + 2\,NaCl$. Both the positive and the negative must be well washed after each process, particularly after the last. The picture is then ready for mounting. In fine portrait work both the negative and the positive are retouched. This consists in removing blemishes with colored pencils or India ink.

The negative.—No. 1. Dissolve: sulphite soda crystals, 2 oz. (57^g) in 8 oz. (236^{cc}) water (distilled); citric acid, 60 grains (4^g) in ½ oz. (15^{cc}) water; bromide ammonium, 25 grains (1½^g) in ½ oz. water; pyrogallic acid, 1 oz. (28^g) in 3 oz. (90^{cc}) water. After dissolving, mix in the order named, and filter. No. 2. Dissolve: sulphite soda, 2 oz. (57^g) in 4 oz. (118^{cc}) water; carbonate potash, 4 oz. (113^g) in 8 oz. (236^{cc}) water. Dissolve separately, mix, and filter. To develop plates, mix 1 dram (3⅔^{cc}) of No. 1 and 1 dram of No. 2 with 2 oz. (60^{cc}) water. Cover the plate with the mixture, and leave as long as the picture increases in distinctness. Remove, wash, and put it into a saturated solution of alum for a minute or two, then wash and put it into a half-saturated solution of hypo. Leave till no white AgCl is seen through the back of the plate. Wash it well.

The positive.—1. Dissolve 30 grains (2^g) pure gold chloride in 15 oz. (450^{cc}) water. This forms a stock solution. 2. Make a saturated solution of borax. 3. Prepare a toning bath by adding ½ oz. (15^{cc}) of the gold chloride solution and 1 oz. (30^{cc}) of the borax solution to 7 oz. (210^{cc}) water. After printing the picture, wash it in 3 or 4 waters, put it into the toning bath, and leave it till considerably darker than desired; wash, and put it for 15 minutes into a hypo solution that has been, after saturation, diluted with 3 or 4 volumes of water. Then wash repeatedly.

CHAPTER LII.

PLATINUM AND GOLD.

PLATINUM.

Examine platinum foil and wire.

285. Platinum is much rarer than gold, and is about two-thirds as costly as the latter. It is found alloyed with other metals, as Au, and is obtained from sand, in which it occurs, by washing. Aqua regia is the only acid which dissolves it, and the action is much slower than with Au. Pt is one of the heaviest metals, having a specific gravity three times that of Fe, or twenty-one and a half times that of water. Its fusing-point is about 1600°, or just below the temperature of the oxy-hydrogen flame. Like Au it has little affinity for other elements, but alloys with many metals. Pt is so tenacious that it can be drawn into wire invisible to the naked eye, being drawn out in the center of a silver wire, which is afterwards dissolved away from the Pt by HNO_3. Noting its valences, 2 and 4, write the symbols for the *ous* and *ic* chlorides and oxides.

286. Uses.—Pt is much used in chemistry in the form of foil, wire, and crucibles. On what properties does this use depend? Describe its use in making H_2SO_4. See page 65.

$PtCl_4$ is made by dissolving Pt in aqua regia, and evaporating the liquid. On heating $PtCl_4$, half of its Cl is given up, leaving $PtCl_2$. If it be still more strongly heated, the Cl all passes off, leaving spongy Pt. By fusing this in the oxy-hydrogen flame, ordinary Pt is obtained. Spongy Pt has a remarkable power of absorbing, or occluding, O without uniting with it. This O it gives up to some other substances, and thus becomes indirectly an oxidizing agent. What other element has this property of occluding gases?

GOLD.

Examine auriferous quartz, gold chloride, yellow and ruby glass colored with gold.

287. Gold is rarely found combined, and has small affinity for other elements, though forming alloys with Cu, Ag, and Hg. Its source is usually either quartz rock, called auriferous quartz, or sand in placer mines. The element is widely distributed, occurring in minute quantities in most soils, sea water, etc. California and Australia are the two greatest gold-producing countries. That from California has a light color, due to a slight admixture of Ag. Australian gold is of a reddish hue, due to an alloy of Cu. Gold-bearing quartz is pulverized, and treated with Hg to dissolve the precious metal, which is then separated from the alloy by distillation. Compare this with the preparation of Ag.

Such is the malleability of Au that it has been hammered into sheets not over one-millionth of an inch thick; it is then as transparent as glass. Gold does not tarnish or change below the melting-point. On account of its softness it is usually alloyed with Cu, sometimes with Ag. Pure gold is twenty-four carats fine. Eighteen carat gold has eighteen parts Au and six Cu. Gold coin has nine parts Au to one part Cu. The most important compound is $AuCl_3$. Describe a use of it. See page 168. This metal is much employed in electroplating, and somewhat in coloring glass.

CHAPTER LIII.

CHEMISTRY OF ROCKS.

288. Classification. — Rocks may be divided, according to their origin, into three classes: (1) Aqueous rocks. These have been formed by deposition of sedimentary material, layer by layer, on the bottoms of ancient oceans, lakes, and rivers, from which they have gradually been raised, to form dry land. (2) Eruptive or volcanic rocks. These have been forced, as hot fluids, through rents and fissures from the interior of the earth. (3) Metamorphic rocks. These, by the combined action of heat, pressure, water, and chemical agents, have been crystallized and chemically altered. The rocks of the first class, such as chalk, limestone, shale, and sandstone, are distinguished by the existence of fossils in them, or by the successive layers of the material which goes to make up their structure and to give them a stratified appearance. The rocks of the second class are recognized by their resemblance to the products of modern volcanoes and their non-stratified appearance. Rocks of the third class are composed of crystals, which, though often very minute, are minerals having a definite chemical composition. Examples of the third class are gneiss, slate, schist, and marble. The last two classes abound on the Eastern sea-board, while the interior of our continent is composed almost exclusively of stratified sedimentary rocks.

289. Composition. — Rocks are not definite compounds, but variable mixtures of minerals. Some, however, are tolerably pure, as limestone ($CaCO_3$) and sandstone.

Granite is mainly made up of three minerals,—quartz, feldspar, and mica. Quartz, when pure, is SiO_2. Feldspar is a mixed silicate of K and Al, and often several other metals, $K_2Al_2Si_6O_{16}$ ($=K_2O$, Al_2O_3, $6\,SiO_2$) symbolizing one variety, while a variety of mica is $H_8Mg_5Fe_7Al_2Si_8O_{18}$.

The pupil should learn to distinguish the different minerals in granite. Quartz is glassy, mica is in scales, usually white or black, and feldspar is the opaque white or red mineral.

290. Importance of Siliceous Rocks. — Slate and schist are also mixed silicates. Pure sandstone is SiO_2, the red variety being colored by iron.

Igneous rocks are always siliceous. Obsidian is a glassy silicate. A mountain of very pure glass, obsidian, two hundred feet high, has lately been found in the Yellowstone region. We see how important Si is, in the compounds SiO_2 and the silicates, as a constituent of the terrestrial crust. Limestone is the only extensive rock from which it is absent. Always combined with O, it is, next to the latter, the most abundant of elements. Silicates of Al, Fe, Ca, K, Na, and Mg are most common, and these metals, in the order given, rank next in abundance.

291. Soils. — Beds of sand, clay, etc., are disintegrated rock. Sand is chiefly SiO_2; clay is decomposed feldspar, slatestone, etc. Soils are composed of these with an added portion of carbonaceous matter from decaying vegetation, which imparts a dark color. The reddish brown hue so often observed in soils and rocks results from ferric salts.

292. Minerals, of which nearly 1000 varieties are now known, may be simple substances, as graphite and sulphur, or compounds, as galena and gypsum. Only seven systems of crystallizations are known, but these are so modified as to give hundreds of forms of crystals. See Physics, page 21. A given chemical substance usually occurs in one system only, but we saw in the case of S that this was not always true.

Crystals of some substances deliquesce, or take water from the air, and thus dissolve themselves. See page 142. Some compounds cannot exist in the crystalline form without a certain percentage of water. This is called "water of crystallization"; if it passes into the air by evaporation, the crystal crumbles to a powder. and is then said to effloresce. See page 139.

293. The Earth's Interior.—We are ignorant of the chemistry of the earth's interior. The deepest boring is but little more than a mile, and volcanic ejections probably come from but a very few miles below the surface. The specific gravity of the interior is known to be more than twice that of the surface rock. From this it has been imagined that towards the center heavy metals like Fe and Au predominate; but this is by no means certain, since the greater pressure at the interior would cause the specific gravity of any substance to increase.

294. Percentage of Elements.—Compute the percentage of O in the following rocks, which compose a large proportion of the earth's crust: SiO_2, Al_2SiO_4, $CaCO_3$. Find the percentage of O in pure water. In air. Taking cellulose, $C_{18}H_{30}O_{15}$, as the basis, find the percentage of O in vegetation.

An estimate, based on Bunsen's analysis of rocks, of the chief elements in the earth's crust, is as follows:—

O, 46 per cent.	Ca, 3 per cent.
Si, 30 "	Na, 2 "
Al, 8 "	K, 2 "
Fe, 6 "	Mg, 1 "

More than half the elements are known to exist in sea water, and the rest are thought to be there, though dissolved in such small quantity as to elude detection. What four are found in the atmosphere?

CHAPTER LIV.

ORGANIC CHEMISTRY.

295. General Considerations. — Inorganic chemistry is the chemistry of minerals, or unorganized bodies. Organic chemistry was formerly defined as the chemistry of the compounds found in plants and animals; but of late it has taken a much wider range, and is now defined as the chemistry of the C compounds, since C is the nucleus around which other elements centre, and with which they combine to form the organic substances. New organic compounds are constantly being discovered and synthesized, so that nearly 100,000 are now known. The molecule of organic matter is often very complex, sometimes containing hundreds of atoms.

In organic as in inorganic chemistry, atoms are bound together by chemical affinity, though it was formerly supposed that an additional or vital force was instrumental in forming organic compounds. For this reason none of these substances, it was thought, could be built up in the laboratory, although many had been analyzed. In 1828 the first organic compound, urea, was artificially prepared, and since then thousands have been synthesized. They are not necessarily manufactured from organic products, but can be made from mineral matter.

296. Molecular Differences. — Molecules may differ in three ways: (1) In the *kind* of atoms they contain. Compare CO_2 and CS_2. (2) In the *number* of atoms.

Compare CO and CO_2. (3) In the *arrangement* of atoms, *i.e.* the molecular structure. Ethyl alcohol and methyl ether have the same number of the same elements, C_2H_6O, but their molecular structure is not the same, and hence their properties differ.

Qualitative analysis shows what elements enter into a compound; quantitative analysis shows the proportion of these elements; structural analysis exhibits molecular structure, and is the branch to which organic chemists are now giving particular attention.

A specialist often works for years to synthesize a series of compounds in the laboratory.

297. Sources. — Some organic products are now made in a purer and cheaper form than Nature herself prepares them. Alizarine, the coloring principle of madder, was until lately obtained only from the root of the madder plant; now it is almost wholly manufactured from coal-tar, and the manufactured article serves its purpose much better than the native product. Ten million dollars' worth is annually made, and Holland, the home of the plant, is giving up madder culture. Artificial naphthol-scarlet is abolishing the culture of the cochineal insect. Indigo has also been synthesized. Certain compounds have been predicted from a theoretical molecular structure, then made, and afterwards found to exist in plants. Others are made that have no known natural existence. The source of a large number of artificial organic products is coal-tar, from bituminous coal. Saccharine, a compound with two hundred and eighty times the sweetening power of sugar, is one of its latest products. Wood, bones, and various fermentable liquids are other sources of organic compounds.

298. Marsh-Gas Series. — The chemistry of the hydrocarbons depends on the valence of C, which, in most cases, is a tetrad. Take successively 1, 2, and 3 C atoms, saturate with H, and note the graphic symbols: —

$$H-\underset{\underset{H}{|}}{\overset{\overset{H}{|}}{C}}-H, \text{ or } CH_4. \qquad H-\underset{\underset{H}{|}}{\overset{\overset{H}{|}}{C}}-\underset{\underset{H}{|}}{\overset{\overset{H}{|}}{C}}-H, \text{ or ?} \qquad H-\underset{\underset{H}{|}}{\overset{\overset{H}{|}}{C}}-\underset{\underset{H}{|}}{\overset{\overset{H}{|}}{C}}-\underset{\underset{H}{|}}{\overset{\overset{H}{|}}{C}}-H, \text{ or ?}$$

Write the graphic and common symbols for 4, 5, and 6 C atoms, saturated with H. Notice that the H atoms are found by doubling the C atoms and adding 2. Hence the general formula for this series would be C_nH_{2n+2}. Write the common symbol for C and H with ten atoms of C; twelve atoms; thirteen. This series is called the marsh-gas series. The first member, CH_4, methäne, or marsh gas, may be written CH_3H, methyl hydride, CH_3 being the methyl radical. C_2H_6, ethäne, the second one, is ethyl hydride, C_2H_5H. Theoretically this series extends without limit; practically it ends with $C_{35}H_{72}$.

In each successive compound of the following list, the C atoms increase by unity. Give the symbols and names of the compounds, and commit the latter to memory: —

					Boiling-point.
1.	CH_4	methäne, or CH_3H,		methyl hydride,	gas.
2.	C_2H_6	ethäne,	C_2H_5H,	ethyl hydride,	"
3.	C_3H_8	propäne,	C_3H_7H,	propyl hydride,	"
4.	?	butäne,	?	?	1°
5.	?	pentäne	?	?	38°
6.	?	hexäne,	?	?	70°
7.	?	heptäne,	?	?	98°
8.	?	octäne,	?	?	125°
9.	?	nonäne,	?	?	148°
10.	?	dekäne,	?	?	171°

Note a successive increase of the boiling-point of the compounds. Crude petroleum contains these hydro-carbons up to 10. Petroleum issues from the earth, and is separated into the different oils by fractional distillation and subsequent treatment with H_2SO_4, etc. Rhigoline is mostly 5 and 6; gasoline, 6 and 7; benzine, 7; naphtha, 7 and 8; kerosene, 9 and 10. Below 10 the compounds are solids. None of those named, however, are pure compounds. Explosions of kerosene are caused by the presence of the lighter hydro-carbons, as naphtha, etc. Notice that, in going down the list, the proportion of C to H becomes much greater, and the lower compounds are the heavy hydro-carbons. To them belong vaseline, paraffine, asphaltum, etc.

299. Alcohols. — The following replacements will show how the symbols for alcohols, ethers, etc., are derived from those of the marsh-gas series. Notice that these symbols also exhibit the molecular structure of the compound. In CH_3H by replacing the last H with the radical OH, we have CH_3OH, methyl hydrate. By a like replacement C_2H_5H becomes C_2H_5OH, ethyl hydrate. These hydrates are alcohols, and are known as methyl alcohol, ethyl alcohol, etc. The common variety is C_2H_5OH. How does this symbol differ from that for water, HOH? Notice in the former the union of a positive, and also of a negative, radical.

Complete the table below, making a series of alcohols, by substitutions as above from the previous table.

1. CH_3OH, methyl hydrate, or methyl alcohol.
2. C_2H_5OH, ethyl hydrate, or ethyl alcohol.
3. ?   ?   ?
4. ?   ?   ?
5. ?   ?   ?

Continue in like manner to 10.

The graphic symbol for CH_3OH is —

$$\begin{array}{c} H \\ | \\ H-C-OH; \\ | \\ H \end{array}$$

for C_2H_5OH it is

$$\begin{array}{c} \text{H} \quad \text{H} \\ | \quad | \\ \text{H}-\text{C}-\text{C}-\text{OH}. \\ | \quad | \\ \text{H} \quad \text{H} \end{array}$$

Write it for the next two.

300. Ethers. — Another interesting class of compounds are the oxides of the marsh-gas series. In this series, O replaces H. CH_3H becomes $(CH_3)_2O$, and C_2H_5H becomes $(C_2H_5)_2O$. Why is a double radical taken? These oxides are ethers, common or sulphuric ether being $(C_2H_5)_2O$. Complete this table, by substituting O in place of H, in the table on page 176.

1. $(CH_3)_2O$, methyl oxide, or methyl ether.
2. $(C_2H_5)_2O$, ethyl oxide, or ethyl ether.
3.   ?   ?   ?
4.   ?   ?   ?
5, etc. ?   ?   ?

Graphically represented the first two are :—

(1) $\quad \begin{array}{c} \text{H} \quad \text{H} \\ | \quad | \\ \text{H}-\text{C}-\text{O}-\text{C}-\text{H}. \\ | \quad | \\ \text{H} \quad \text{H} \end{array}$
(2) $\quad \begin{array}{c} \text{H} \quad \text{H} \quad\quad \text{H} \quad \text{H} \\ | \quad | \quad\quad | \quad | \\ \text{H}-\text{C}-\text{C}-\text{O}-\text{C}-\text{C}-\text{H}. \\ | \quad | \quad\quad | \quad | \\ \text{H} \quad \text{H} \quad\quad \text{H} \quad \text{H} \end{array}$

301. Substitutions. — A large number of other substitutions can be made in each symbol, thus giving rise to as many different compounds.

In CH_4, by substituting 3 Cl for 3 H, —

$$\begin{array}{c} \text{H} \\ | \\ \text{H}-\text{C}-\text{H} \\ | \\ \text{H} \end{array} \text{becomes} \begin{array}{c} \text{Cl} \\ | \\ \text{H}-\text{C}-\text{Cl}, \\ | \\ \text{Cl} \end{array} \text{or } CHCl_3, \text{ the symbol for chloroform.}$$

Replace successively one, two, and four atoms with Cl, and write the common symbols. Make the same changes with Br. For each atom of H in CH_4 substitute the radical CH_3, giving the graphic and common formulæ. Also substitute C_2H_5. Are these radicals positive or negative?

From the above series of formulæ, of which CH_4 is the basis, are derived, in addition to the alcohols and ethers, the natural oils, fatty acids, etc.

302. Olefines. — A second series of hydro-carbons is represented by the general formula C_nH_{2n}. The first member of this series is C_2H_4, or, graphically, —

$$\begin{array}{cc} H & H \\ | & | \\ C & = C. \\ | & | \\ H & H \end{array}$$

Compare it with that for C_2H_6, in the first series, noting the apparent molecular structure of each.

$$\begin{array}{ccc} H & & H \\ | & & | \\ C = C & - & C - H, \text{ or } C_3H_6 \text{ is the second member.} \\ | & | & | \\ H & H & H \end{array}$$

Write formulæ for the third and fourth members.

Write the common formulæ for the first ten of this series. This is the olefiant-gas series, and to it belong oxalic and tartaric acids, glycerin, and a vast number of other compounds, many of which are derived by replacements.

303. Other Series. — In addition to the two series of hydro-carbons above given, C_nH_{2n+2} and C_nH_{2n}, other series are known with the general formulæ C_nH_{2n-2}, C_nH_{2n-4}, C_nH_{2n-6}, C_nH_{2n-8}, etc., as far as C_nH_{2n-32}, or $C_{20}H_{20}$. Each of these has a large number of representatives, as was found in the marsh-gas series. Not far from two hundred direct compounds of C and H are known, not to mention substitutions. The formula C_nH_{2n-6} represents a large and interesting group of compounds, called the benzine series. This is the basis of the aniline dyes, and of many perfumes and flavors.

CHAPTER LV.

ILLUMINATING GAS.

304. Source. — The three main elements in combustion are O, H, C. Air supplies O, the supporter; C and H are usually united, as hydro-carbons, in luminants and combustibles. H gives little light in burning; C gives much. The fibers of plants contain hydro-carbons, and by destructive distillation these are separated, as gases, from wood and coal, and used for illuminating purposes. Mineral coal is fossilized vegetable matter; anthracite has had most of the volatile hydro-carbons removed by distillation in the earth; bituminous and cannel coals retain them. These latter coals are distilled, and furnish us illuminating gas.

Experiment 129. — Put into a t.t. 20^g of cannel coal in fine pieces. Heat, and collect the gas over H_2O. Test its combustibility. Notice any impurities, such as tar, adhering to the sides of the t.t., or of the receiver after combustion. Try to ignite a piece of cannel coal by holding it in a Bunsen flame. Is it the C which burns, or the hydro-carbons? Distil some wood shavings in a small ignition-tube, and light the escaping gas.

305. Preparation and Purification. — To make illuminating gas, fire-clay retorts filled with coal are heated to 1100° or more, over a fire of coke or coal. Tubes lead the distilled gas into a horizontal pipe, called the hydraulic main, partly filled with water, into which the ends of the gas-pipe dip. The gas then passes through condensers consisting of several hundred feet of vertical pipe, through high towers, called washers, in which a fine spray

ILLUMINATING GAS. 181

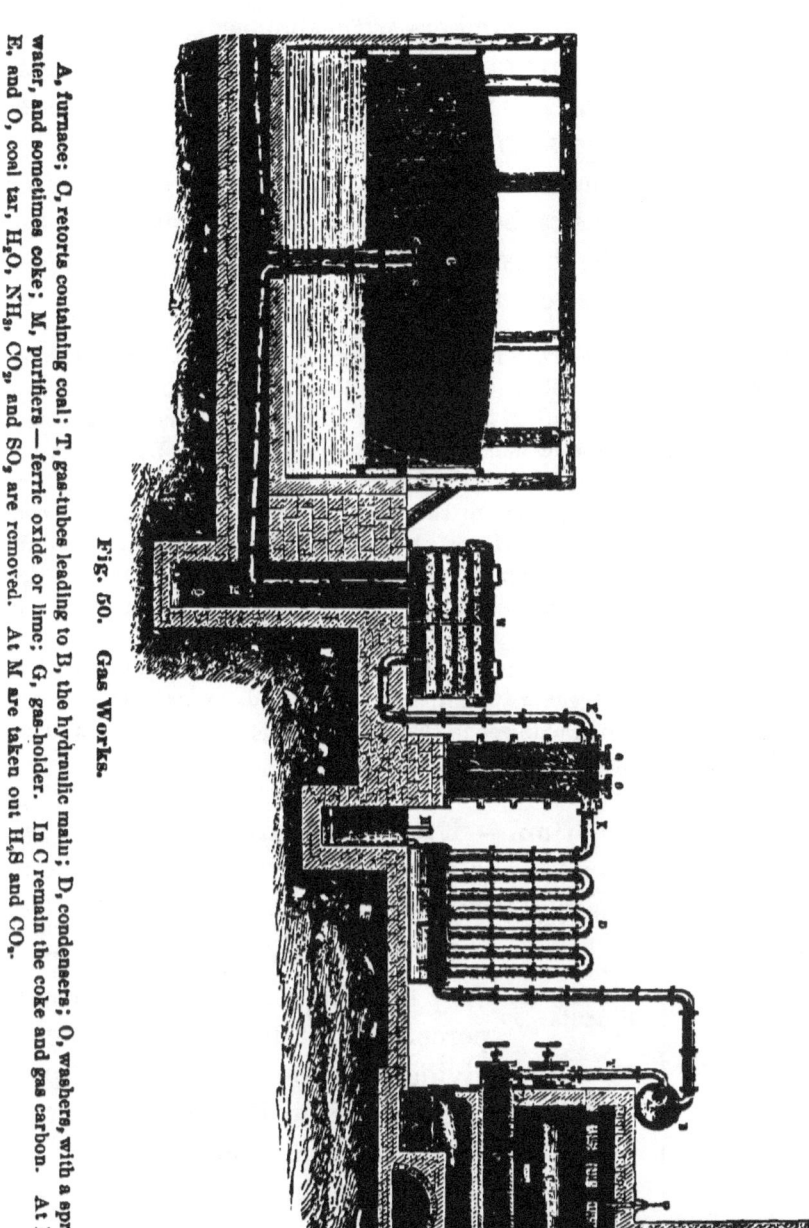

Fig. 50. Gas Works.

A, furnace; C, retorts containing coal; T, gas-tubes leading to B, the hydraulic main; D, condensers; O, washers, with a spray of water, and sometimes coke; M, purifiers — ferric oxide or lime; G, gas-holder. In C remain the coke and gas carbon. At B, D, E, and O, coal tar, H_2O, NH_3, CO_2, and SO_2 are removed. At M are taken out H_2S and CO_2.

of water falls, into chambers with shelves containing the purifiers CaO or hydrated Fe_2O_3, and finally into a gas-holder, whence it is distributed. At the hydraulic main, condensers, washers, and purifiers, certain impurities are removed from the gas. Coke is the solid C residue after distillation. Gas-carbon, also a solid, is formed by the separation of the heavier hydro-carbons at high temperature, and is deposited on the sides of the retort.

Coal gas, as it leaves the retort, has many impurities. It is accompanied with about $\frac{1}{8}$ its weight of coal tar, $\frac{1}{2}$ its weight of H_2O vapor, $\frac{1}{50}$ NH_3, $\frac{1}{20}$ CO_2, $\frac{1}{20}$ to $\frac{1}{50}$ H_2S, $\frac{1}{300}$ to $\frac{1}{500}$ S in other forms. The tar is mostly taken out at the hydraulic main, which also withdraws some H_2O with other impurities in solution. The condensers remove the rest of the tar, and the H_2O, except what is necessary to saturate the gas. At the main, the condensers, and the washers, NH_3 is abstracted, CO_2 and H_2S are much reduced, and the other S compounds are diminished. Lime purification removes CO_2 and H_2S, and, to some extent, other S compounds. Iron purification removes H_2S. $Fe_2O_3 + 3 H_2S = 2 FeS + S + 3 H_2O$.

The FeS is revivified by exposure to the air. $2 FeS + O_3 = Fe_2O_3 + 2 S$. It can then be used again. H_2S, if not separated, burns with the gas, forming H_2SO_3, which oxidizes in the air to H_2SO_4; hence the need of removing it. CO_2 diminishes the illuminating power.

306. Composition. — Even when freed from its impurities coal-gas is a very complex mixture, the chief components being nearly as follows: —

	Per cent.	
H	45.	⎫
CH_4	41.	⎬ diluents.
CO	5.	⎭
C_6H_6	1.3	⎫
C_3H_6	1.2	⎬ luminants.
C_2H_4	2.5	⎭
CO_2	2.	⎫ impurities.
N, etc.	2.	⎭
	100	

Diluents, having little C, give very little light. Notice the small percentage of luminants, or light-giving compounds, also the proportion of C to H in them.

Cannel coal contains more of the heavy hydro-carbons, C_nH_{2n}, etc., than the ordinary bituminous coal. Ten per cent of the coal should be cannel; naphtha is, however, often employed to subserve the same purpose, one ton of ordinary bituminous coal requiring four gallons of oil.

In Boston, 7,000,000 cubic feet of gas have been burned in one day, consuming 500 tons of coal; the average is not more than half that quantity. Of the other products, coke is employed for heating purposes, gas carbon is used to some extent in electrical work, and coal-tar is the source of very many artificial products that were formerly only of natural origin. NH_3 is the main source of ammonium salts, and S is made into H_2SO_4.

307. Natural Gas occurs near Pittsburg, Pa., and in many other places, in immense quantities. It is not only employed to light the streets and houses, but is used for fires and in iron and glass manufactories. It is estimated that 600,000,000 cubic feet are burned, saving 10,000 tons of coal daily in Pittsburg. Only half a dozen factories now use coal. More than half the gas is wasted through safety valves, on account of the great pressure on the pipes as it issues from the earth.

These reservoirs of natural gas very frequently occur in sandstone, usually in the vicinity of coal-beds, but sometimes remote from them. In all cases the origin of the gas is thought to be in the destructive distillation, extending through long geological periods, of coal or of other vegetable or animal matter in the earth's interior.

Natural gas varies in composition, and even in the same well, from day to day; it consists chiefly of CH_4, with some other hydro-carbons.

CHAPTER LVI.

ALCOHOL.

308. Fermented Liquor.

Experiment 130. — Introduce 20^{cc} of molasses into a flask of 200^{cc}, fill it with water to the neck, and put in half a cake of yeast. Fit to this a d.t., and pass the end of it into a t.t. holding a clear solution of lime water. Leave in a warm place for two or three days. Then look for a turbidity in the lime water, and account for it (page 79). See whether the liquid in the flask is sweet. The sugar should be changed to alcohol and CO_2. This is fermented liquor; it contains a small percentage of alcohol.

309. Distilled Liquor.

Experiment 131. — Attach the flask used in the last experiment to the apparatus for distilling water (Fig. 32), and distil not more than one-fifth of the liquid, leaving the rest in the flask. The greater part of the alcohol will pass over. To obtain it all, at least half of the liquid must be distilled; what passes over towards the last is mostly water. Taste and smell the distillate. Put some into an e.d. and touch a lighted match to it. If it does not burn, redistil half of the distillate and try to ignite the product. Try the combustibility of commercial alcohol; of Jamaica ginger, or of any other liquid known to contain alcohol.

310. Effect on the System.

Experiment 132. — Put a little of the white of egg into an e.d. or a beaker; cover it with strong alcohol and note the effect. Strong alcohol has the same coagulating action on the brain and on the tissues generally, when taken into the system, absorbing water from them, hardening them, and contracting them in bulk.

311. Affinity for Water.

Experiment 133. — To show the contraction in mixing alcohol and water, measure exactly 5^{cc} of alcohol and 5^{cc} of water. Pour them together, and presently measure the mixture. The volume is diminished. A strip of parchment soaked in water till it is limp, then dipped into strong alcohol, becomes again stiff, owing to the attraction of alcohol for water.

312. Purity. — The most important alcohols are methyl alcohol and ethyl alcohol. The former, wood spirit, is obtained in an impure state by distilling wood; it is used to dissolve resins, fats, oils, etc., and to make aniline. It is poisonous, as are the others.

Ethyl alcohol, spirit of wine, is the commercial article. It is prepared by fermenting glucose, and distilling the product. It boils at 78°, vaporizing 22° lower than water, from which it can be separated by fractional distillation. By successive distillations of alcohol ninety-four per cent can be obtained, which is the best commercial article, though most grades fall far below this. Five per cent more can be removed by distilling with CaO, which has a strong affinity for water. The last one per cent is removed by BaO. One hundred per cent constitutes absolute alcohol, which is a deadly poison. Diluted, it increases the circulation, stimulates the system, hardens the tissues by withdrawing water, and is the intoxicating principle in all liquors. It is very inflammable, giving little light, and much heat, and readily evaporates.

Beer has usually three to six per cent of alcohol; wines, eight to twenty per cent. The courts now regard all liquors having three per cent, or less, of alcohol, as not intoxicating. In Massachusetts it is one per cent.

CHAPTER LVII.

OILS, FATS, AND SOAPS.

313. Sources and Kinds of Oils and Fats. — Oils and fats are insoluble in water; the former are liquid, the latter solid. Most fats are obtained from animals, oils from both plants and animals. Oils are classified as *fixed* and *essential*. Castor oil is an example of the former and oil of cloves of the latter. Fixed oils include *drying* and *non-drying* oils. They leave a stain on paper, while essential, or volatile oils, leave no trace, but evaporate readily. Essential oils dissolved in alcohol furnish essences. They are obtained by distilling with water the leaves, petals, etc., of plants. Drying oils, as linseed, absorb O from the air, and thus solidify. Non-drying ones, as olive, do not solidify, but develop acids and become rancid after some time.

Oils and fats are salts of fatty acids and the base glycerin. The three most common of these salts are oleïn, found in olive oil, palmitin, in palm oil and human fat, and stearin, in lard. The first is liquid, the second semi-solid, the last solid. Most fats are mixtures of these and other salts.

Oleïn	=	Glyceryl oleate	⎫	oleïc	⎫	
Palmitin	=	Glyceryl palmitate	⎬ salts from	palmitic	⎬ acid, and	glyceryl hydrate.
Stearin	=	Glyceryl stearate	⎭	stearic	⎭	

314. Saponification consists in separating these salts into their acids and the base glycerin; soap-making is the best illustration. To effect this separation, a strong soluble base is used, KOH for soft, and NaOH for hard soap. Study this reaction:

$$\left.\begin{array}{l}\text{Glyceryl oleate}\\ \text{Glyceryl palmitate}\\ \text{Glyceryl stearate}\end{array}\right\} + \left\{\begin{array}{l}\text{sodium}\\ \text{hydrate}\end{array}\right\} = \text{sodium} \left\{\begin{array}{l}\text{oleate}\\ \text{palmitate}\\ \text{stearate}\end{array}\right\} + \left\{\begin{array}{l}\text{glyceryl}\\ \text{hydrate.}\end{array}\right.$$

Soaps are thus salts of fatty acids and of K or Na.

315. Soap is soluble in soft water, but the sodium stearate probably unites with water to form hydrogen sodium stearate and NaOH. The grease which exudes from the skin, or appears in fabrics to be washed, is attacked by this NaOH and removed, together with the suspended dirt, and a new soap is formed and dissolved in the water. Hard water contains salts of Ca and Mg, and when soap is used with it the Na is at once replaced by these metals, and insoluble Ca or Mg soaps are formed. Hence in hard water soap will not cleanse till all the Ca and Mg compounds have combined.

316. Glycerin, $C_3H_5(OH)_3$, is a sweet, thick, colorless, unctuous liquid, used in cosmetics, unguents, pomades, etc. It is prepared in quantity by passing superheated steam over fats when under pressure.

317. Dynamite. — Treated with HNO_3 and H_2SO_4 glycerin forms the very explosive and poisonous liquid nitro-glycerin. In this process the $C_3H_5(OH)_3$ becomes $C_3H_5(NO_3)_3$. $C_3H_5(OH)_3 + 3\ HNO_3 = C_3H_5(NO_3)_3 + 3\ H_2O$. H_2SO_4 is used to absorb the H_2O which is formed. Nitro-glycerin, absorbed by gunpowder, diatomaceous earth, sawdust, etc., forms dynamite. For obvious reasons the pupil should not experiment with these substances.

318. Butter and Oleomargärine. — Milk contains minute particles of fat, about $\frac{1}{500}$ of an inch in diameter, which give it the white color. These particles are lighter than the containing liquid, and rise to the top as cream. Churning unites the particles more closely, and separates them from the buttermilk. The flavor of butter is due to the presence of five or ten per cent of butyric and other acids of the same series.

It was found that cows gave milk after they ceased to have food; hence it was inferred that the milk was produced at the expense of the cows' fat. Why could not butter be artificially made from the same fat? It was but a step from fat to milk, as it was from milk to butter. Oleomargärine, or butterine, was the result. Beef fat, suet, is washed in water, ground to a pulp, and partially melted and strained, the stearin is separated from the filtered liquid and made into soap, and an oily liquid is left. This is salted, colored with annotto, mixed with a certain portion of milk, and churned. The product is scarcely distinguishable from butter, and is chemically nearly identical with it, though less likely to become rancid from the absence of certain fatty acids; its cost is perhaps one-third as much as that of butter.

CHAPTER LVIII.

CARBO-HYDRATES.

319. Carbon and Water. — Some very important organic compounds have H and O, in the proper proportion to form water, united with C   The three leading ones are sugar, $C_{12}H_{22}O_{11}$ or $C_{12}(H_2O)_{11}$, starch, $C_6H_{10}O_5$ or ?, and cellulose, $C_{18}H_{30}O_{15}$ or ?. Note the significance of the name carbo-hydrates as applied to them.

320. Sugars may be divided into two classes, — the sucroses, $C_{12}H_{22}O_{11}$, and the glucoses, $C_6H_{12}O_6$. Sucrose, the principal member of the first class, is obtained from the juice of the maple, the palm, the beet and the sugar-cane; in Europe largely from the beet, in America from cane. Granulated sugar is that which has been refined; brown sugar is the unrefined. From the sap evaporated by boiling, brown sugar crystallizes, leaving molasses, which contains glucose and other substances. Good molasses has but a small percentage of glucose. To refine brown sugar it is dissolved in water, a small quantity of blood is added to remove certain vegetable substances, after which it is filtered through animal charcoal, *i.e.* bone-black, a process which takes out the coloring-matter. The water is then evaporated in vacuum-pans, so as to boil at about 74° and to prevent conversion into grape sugar. By this process much glucose or syrup is formed, which is separated from the crystalline sucrose by rapidly revolving centrifugal machines.

Great quantities of sucrose are used for food by all civilized nations. A single refinery in New York purifies 2,000,000 pounds per day.

321. Glucose, or invert sugar, the principal member of the second class, consists of two distinct kinds of sugar, — dextrose and levulose. These differ in certain properties, but have the same symbol. Both are found in equal parts in ripe fruits, while sucrose occurs in the unripe. Honey contains these three kinds of sugar.

Sucrose, by the action of heat, weak acids, or ferments, may be resolved into the other two varieties. $C_{12}H_{22}O_{11} + H_2O = C_6H_{12}O_6 + C_6H_{12}O_6$. No mode of reversing this process, or of transforming glucose into sucrose is known. Glucose is easily made from starch or from the cellulose in cotton rags, sawdust, etc. If boiled with dilute H_2SO_4 starch takes up water and becomes glucose. $C_6H_{10}O_5 + H_2O = C_6H_{12}O_6$.

$CaCO_3$ is added to precipitate the H_2SO_4, which remains unchanged. State the reaction. The product is filtered and the filtrate is evaporated. Much glucose is made from the starch of corn and potatoes.

322. Starch is found in all plants, especially in grains, seeds, and tubers. Green plants — those containing chlorophyll — manufacture their own starch from CO_2 and H_2O. These chlorophyll grains are the plant's chemical laboratories, and hundreds of thousands of them exist in every leaf. CO_2 and a very little H_2O enter the leaf from the air, H_2O being also drawn up through the root and stem from the earth. In some unknown way in the leaf, light has the power of synthesizing these into starch and setting free O, which is returned to the atmosphere.

$6\ CO_2 + 5\ H_2O = C_6H_{10}O_5 + 12\ O$. As no such change takes place in darkness, all green plants must have light. Parasitic plants, which are usually colorless, obtain starch ready-made from those on which they feed.

323. Uses. — Glucose is used in the manufacture of alcohol and cheap confectionery, and in adulterating sucrose. It is only two-thirds as sweet as the latter. The seeds of all plants contain starch for the germinating sprout to feed upon; but starch is insoluble, and hence useless until it is converted into glucose. This is effected by the action of warmth, moisture, and a ferment in the seed. Glucose is soluble and is at first the plant's main food.

Commercial starch is made in the United States chiefly from corn; in Europe, from potatoes. Differences in the size of starch granules enable microscopists to determine the plant to which they belong.

324. Cellulose, or woody fiber, is the basis of all vegetable cell walls. Cotton fiber represents almost pure cellulose. From it are made paper and woven tissues. In paper manufacture, woody fiber is made into a pulp, washed, bleached, filtered, hot-pressed, and sometimes glazed. Parchment paper, vegetable parchment, is made by dipping unglazed paper for half a minute into cold dilute H_2SO_4, 1 part H_2O, $2\frac{1}{2}$ parts H_2SO_4, and then washing. The fiber, by chemical change, is thus toughened. The cell walls of wood are impure cellulose; hence the inferior quality of paper made from wood-pulp. Paper is now employed for a large number of purposes for which wood has heretofore been used, such as for barrels, pails, and other hollow ware, wheels, etc.

325. Gun-cotton is made by treating cotton fiber with H_2SO_4 and HNO_3, washing and drying. To all appearances no change has taken place, but the substance has become an explosive compound.

326. Dextrin, a gummy substance used for the backs of postage stamps, is a carbo-hydrate, as in fact are gums in general. Dextrin is made by heating starch with H_2SO_4 at a lower temperature than for dextrose.

327. Zylonite and Celluloid. — These two similar substances embody the latest use of cellulose in manufactured articles. For zylonite, linen paper is cut into strips two feet by one inch, soaked ten minutes in a mixture of H_2SO_4 and HNO_3, a process called nitration, washed for several hours, then ground to a fine pulp, and thoroughly dried. It is then similar to pyroxiline. Aniline coloring-matter of any desired shade is added, after which it is dissolved by soaking some hours in alcohol and camphor, the liquid is evaporated, and the substance is kneaded between steam-heated iron rollers, dried with hot air, and finally subjected to great pressure, to harden it, and cut into sheets. Zylonite is combustible at a low temperature, and when in the pyroxiline stage, explosively so. Ivory, coral, amber, bone, tortoise shell, malachite, etc., are so closely imitated that the imitation can only be detected by analysis. Collars, combs, canes, piano-keys, and jewelry, are manufactured from it, and it can be made transparent enough for windows.

CHAPTER LIX.

CHEMISTRY OF FERMENTATION.

328. Ferments. — A large number of chemical changes are brought about through the direct agency of bodies called ferments; their action is called fermentation. Ferments are sometimes lifeless chemical products found in living bodies; but in other cases they are humble plants.

329. Yeast is one of the most common of living ferments, wild yeast being a microscopic plant found on the ground near apple-trees and grape-vines, and often in the air. The cultivated variety is sold by grocers. The temperature best suited to the rapid multiplication of the germs forming the ferment plant is 25° to 35°.

330. Alcoholic and Acetic Fermentation. — The changes which the juice of the apple undergoes in forming cider and vinegar are a good illustration of fermentation by a living plant. Apple-juice contains sucrose. Yeast germs from the air, getting into this unfermented liquor, cause it to "work." This process changes sucrose to glucose, and glucose to alcohol and CO_2, and is known as alcoholic fermentation. The latter reaction, $C_6H_{12}O_6 = 2\,C_2H_6O + 2\,CO_2$, is only partially correct, as other products are formed. The juice has now become cider; the sugar, alcohol. After a time, if left exposed, another organism finds its way to the alcohol, and transforms it into acetic acid, $HC_2H_3O_2$, and H_2O. This process is called acetic fermentation. $C_2H_6O + O_2 = HC_2H_3O_2 + H_2O$. For this fermentation, a liquor should not have over ten per cent of alcohol. Mother of vinegar consists of the germs that caused the fermentation. Still a third species of ferment may cause another action, changing acetic acid to H_2O and CO_2. The vinegar then tastes flat. $HC_2H_3O_2 + 4\ O = 2\ H_2O + 2\ CO_2$.

Some mineral acids, as H_2SO_4 and HCl, and some organic acids, are regarded as lifeless ferments. To this class are thought to belong the diastase of malt and the pepsin of the stomach. This variety of ferments exists in the seeds of all plants, and changes starch to glucose.

331. Bread which is raised by yeast is fermented, the object being to produce CO_2, bubbles of which, with the alcohol, cause the dough to rise and make the bread light.

Grapes and other fruits ferment and produce wines, etc., from which distilled liquors are obtained.

332. Lactic Fermentation changes the sugar of milk, lactose, to lactic acid, *i.e.* sour milk. In canning fruit, any germs present are killed by heating, and those from the air are excluded by sealing the can. Milk has been kept sweet for years by boiling, and tightly covering the receptacle with two or three folds of cotton cloth.

333. Putrefaction is fermentation in which the products of decay are ill-smelling. Saprophytes attack the dead matter, feed on it, and cause it to putrefy. This action, as well as that of ordinary fermentation, used to be attributed solely to oxygen. Germs bring back organic matter to a more elementary state, and so have a very important function. By some scientists, digestion is regarded as a species of fermentation, probably due to the action of lifeless ferments; *e.g.* sucrose cannot be taken into the system, but is first fermented to glucose.

334. Most Infectious Diseases are now thought to be due to parasites of various kinds, such as bacteria, microbes, etc., with which the victim often swarms, and which feed on his tissues, multiplying with enormous rapidity. Such diseases are small-pox, intermittent and yellow fevers, etc. Consumption, or tuberculosis, is believed to be caused by a microbe which destroys the lungs. In some diseases not less than fifteen billions of the organisms are estimated to exist in a cubic inch. These multiply so rapidly that from a single germ in forty-eight hours may be produced nearly three hundred billions. These germs do not spring into life spontaneously from inorganic matter, but come from pre-existent similar forms. Parasites are not so rare in the system even of a healthy person as is generally supposed. They are found on our teeth and in many of the tissues of the body.

Several infectious diseases are now warded off or rendered less virulent by vaccination, the philosophy of which is that the organisms are rendered less dangerous by domestication; several crops, or generations, are grown in a prepared liquid, each less injurious than its parent. Some of the more domesticated ones are introduced into the system, and the person has only a modified form of the disease, often scarcely any at all, and is for a more or less limited time insured against further danger.

Dust particles and motes floating in the air are in part germs, living or dead, often requiring only moisture and mild temperature for resuscitation. Most of these are harmless.

CHAPTER LX.

CHEMISTRY OF LIFE.

335. Growth. — The chemistry of organic life is very complex, and not well understood. A few of the principal points of distinction between the two great classes of living organisms, plants and animals, are all that can be noted here. Minerals grow by accretion, *i.e.* by the external addition of molecules of the same material as their interior. A crystal of quartz grows by the addition of successive molecules of SiO_2, arranged in a symmetrical manner around its axis. The growth of crystals can be seen by suspending a string in a saturated solution of $CuSO_4$, or of sugar. In plants and animals the growth is very much more complex, but is from the interior, and is produced by the multiplication of cells. To produce this cell-growth and multiplication, food-materials must be furnished and assimilated. In plants, sap serves to carry the food-materials to the parts where they are needed. In the higher animals, various fluids, the most important of which is the blood, serve the same purpose.

336. Chemistry of Plants. — In ultimate analysis, plants consist mainly of C, H, O, N, P, K. In proximate analysis, as it is called, they are found to contain these elements combined to form substances like starch, sugar, etc. Water is the leading compound in both animals and plants. One of the most important differences between animals and plants is, that all plants, except parasitic ones, are capable of building up such compounds as starch from mineral food-stuffs, while animals have not that power, but must have the products of proximate analysis ready prepared, as it were, by the plant. Hence plants thrive on minerals, whereas animals feed on plants or on other animals. The power which plants have of transforming mineral matter is largely due to sunlight, the action of which in separating CO_2 was described on page 82. The reaction in the synthesis of starch from CO_2 and H_2O in the leaf, is thought to be as follows: $6\ CO_2 + 5\ H_2O = C_6H_{10}O_5 + 12\ O$. $C_6H_{10}O_5$ is taken into the tree as starch; 12 O is given back to the air. All the constituents, except CO_2 and a very small quantity of H_2O, are absorbed by the roots, from the soil, from which they are soon withdrawn by vegetation. To renew the supply, fertilizers or manures are applied to the soil. These must contain compounds of N, P, and K. N is usually applied in the form of ammonium compounds, *e.g.* $(NH_4)_2SO_4$, $(NH_4)_2CO_3$, and NH_4NO_3. The reduction and application of $Ca_3(PO_4)_2$ for this purpose was described on page 123. K is usually applied in the form of KCl and K_2SO_4.

337. Food of Man. — In the higher animals the object is not so much to increase the size as to supply the waste of the system. The principal elements in man's body are C, H, O, N, S, P.

An illustration of the transformation of mineral foods by plants before they can be used by animals is found in the $Ca_3(PO_4)_2$ of bones. This is rendered soluble; plants absorb and transform it; animals eat the plants and obtain the phosphates. Thus man is said to "eat his own bones." The food of mankind may be divided into four classes: (1) proteids, which contain C, H, O, N, and often S and P; (2) fats, and (3) amyloids, both of which contain C,

H, O; (4) minerals. Examples of the first class are the gluten of flour, the albumen of the white of egg, and the casein of cheese. To the second class belong fats and oils; to the third, starch, sugar, and gums; to the fourth, H_2O, $NaCl$ and other salts. Since only proteids contain all the requisite elements, they are essential to human food, and are the only absolutely essential ones, except minerals; but since they do not contain all the elements in the proportion needed by the system, a mixed diet is indispensable. Milk, better than any other single food, supplies the needs of the system. The digestion and assimilation of these food-stuffs and the composition of the various tissues is too complicated to be taken up here; for their discussion the reader is referred to works on physiological chemistry.

338. Conservation. — Plants, in growing, decompose CO_2, and thereby store up energy, the energy derived from the light and heat of the sun. When they decay, or are burned, or are eaten by animals, exactly the same amount of energy is liberated, or changed from potential to kinetic, and the same amount of CO_2 is restored to the air. The tree that took a hundred years to complete its growth may be burned in an hour, or be many years in decaying; but in either case it gives back to its mother Nature, all the matter and energy that it originally borrowed. The ash from burning plants represents the earthy matter, or salts, which the plant assimilated during its growth; the rest is volatile. In the growth and destruction of plants or of animals, both energy and matter have undergone transformation. Animals, in feeding on plants, transform the energy of sunlight into the energy of vitality. Thus "we are children of the sun."

CHAPTER LXI.

THEORIES.

339. The La Place Theory. — This theory supposes that at one time the earth and the other planets, together with the sun, constituted a single mass of vapor, extending billions of miles in space; that it rotated around its center; that it gradually shrank in volume by the transformation of potential into kinetic energy; that portions of its outer rim were thrown off, and finally condensed into planets; that our sun is only the remainder of that central mass which still rotates and carries the planets around with it; that the earth is a cooling globe; that the other planets are going through the same phases as the earth; and finally that the sun itself is destined like them to become a cold body.

340. A Cooling Earth. — The sun's temperature is variously estimated at many thousands, or even millions of degrees. Many metals which exist on the earth as solids — *e.g.* iron — are gases in the dense atmosphere of the sun. Thus the earth, in its early existence, must have been composed of gases only, which in after ages condensed into liquids and solids.

So intense was the heat at that time, that substances probably existed as elements instead of compounds, *i.e.* the temperature was above the point of dissociation. We have seen that Al_2O_3, CaO, SiO_2, etc., are dissociated at the highest temperatures only. If the temperature were above that of combination, compounds could not exist as such, but matter would exist in its elemental state. On slowly cooling, these elements would combine. It is, then, a fair inference that such compounds as need the highest temperatures to separate them, as silica, silicates, and some oxides, were formed from their elements at a much earlier stage of the earth's history than were those compounds that are more easily separable, such as water, lead sulphide, etc., and that the most infusible substances were solidified first.

341. Evolution. — As the earth slowly cooled, elements united to form compounds, gases condensed to liquids, and these to solids. At one time the entire surface of our planet may have been liquid. When the cooling surface reached a point somewhat below that of boiling water, the lowest forms of life appeared in the ocean. This was many millions of years ago. Most scientists believe that all vegetable and animal life has developed from the lowest forms of life. There is also a theory that all chemical elements are derivatives of hydrogen, or of some other element, and that all the so-called elements are really compounds, which a sufficiently high temperature would dissociate. As evidence of this, it is said that less than half as many elements have been discovered in the sun as in the earth, and that comets and nebulæ, which are less developed forms of matter than the sun, have a few simple substances only.

It is easy to fancy that all living bodies, both animal and vegetable, are only natural growths from the lowest forms of life; that these lowest forms are a development, with new manifestations of energy, from inorganic matter; that compounds are derived from elements; and that the last are derivatives of some one element; but it must be borne in mind that this is only a theory.

342. New Theory of Chemistry. — We have seen that heat lies at the basis of chemical as well as of physical changes. By the loss of heat, or perhaps by the change of potential into kinetic energy, in a nebulous parent mass, planets were formed, capable of supporting living organisms. Heat changes solids to liquids, and liquids to gases; it resolves compounds, or it aids chemical union. In every chemical combination heat is developed; in every case of dissociation heat is absorbed. Properly written, every equation should be: $a + b = c + $ heat; *e.g.* $2H + O = H_2O + $ heat; or, $c - a = b - $ heat; *e.g.* $H_2O - 2H = O - $ heat. Another illustration is the combination of C and O, and the dissociation of CO_2, as given on page 82. $C + O_2 = CO_2 + $ energy. $CO_2 - O_2 = C - $ energy. In fact, there are indications that the present theory of atoms and molecules of matter, as the foundation of chemistry, will at no distant day give place to a theory of chemistry based on the forms of energy, of which heat is a manifestation.

CHAPTER LXII.

GAS VOLUMES AND WEIGHTS.

343. Oxygen.

Experiment 134. — Weigh accurately, using delicate balances, 5^g $KClO_3$, and mix with the crystals 1 or 2^g of pure powdered MnO_2. Put the mixture into a t.t. with a tight-fitting cork and delivery-tube, and invert over the water-pan, to collect the gas, a flask of at least one and a half liters' capacity, filled with water. Apply heat, and, without rejecting any of the gas, collect it as long as any will separate. Then press the flask down into the water till the level in the flask is the same as that outside, and remove the flask, leaving in the bottom all the water that is not displaced. Weigh the flask with the water it contains; then completely fill it with water and weigh again. Substract the first weight from the second, and the result will evidently be the weight of water that occupies the same volume as the O collected. This weight, if expressed in grams, represents approximately the number of cubic centimeters of water, — since 1^{cc} of water weighs 1^g, — or the number of cubic centimeters of O.

At the time the experiment is performed the temperature should be noted with a centigrade thermometer, and the atmospheric pressure with a barometer graduated to millimeters.

Suppose that we have obtained 1450^{cc} of O, that the temperature is $27°$, and the pressure 758^{mm}; we wish to find the volume and the weight of the gas at $0°$ and 760^{mm}.

According to the law of Charles — Physics, page 132 — the volume of a given quantity of gas at constant pressure varies directly as the absolute temperature. To reduce from the centigrade to the absolute scale, we have only to add $273°$. Adding the observed temperature, we have $273° + 27° = 300°$. Applying the above law to O obtained at $300°$ A, we have the proportion below. Since the volume of O at $273°$ will be less than it will at $300°$, the fourth term, or answer will be less than the third, and the second term must be less than the first. $300 : 273 :: 1450 : x$. This would give the result dependent upon temperature alone.

By the law of Mariotte — Physics, page 40 — the volume of a given quantity of gas at a constant temperature varies inversely as the pressure. Applying this law to the O obtained at 758^{mm}, we have the following proportion. The volume at 760^{mm} will be less than at 758^{mm}; or the fourth term will be less than the third; hence the second must be less than the first. $760 : 758 :: 1450 : x$. This would give the result dependent on pressure alone.

Combining the two proportions in one: —

$$\left.\begin{array}{c} 300 : 273 \\ 760 : 758 \end{array}\right\} :: 1450 : x = 1316^{cc}.$$

$1316^{cc} = 1.316^{l}$. It remains to find the weight of this gas. A liter of H weighs 0.0896^{g}. The vapor density of O is 16. Hence 1.316^{l} O will weigh $1.316 \times 16 \times 0.0896 = 1.89^{g}$.

From the equation $\left\{\begin{array}{cc} KClO_3 = KCl + O_3 \\ 122.5 \quad\quad 48 \\ 5 \quad\quad x \end{array}\right\}$ we make a proportion,

$122.5 : 5 :: 48 : x = 1.95$, and obtain, as the weight of O contained in 5^g of $KClO_3$, 1.95^g. The weight we actually obtained was 1.89^g. This leaves an error of 0.06^g, or a little over 4 per cent of error ($0.06 \div 1.95 = 0.03 +$). The percentage of error, in performing this experiment, should fall within 10.

Some of the liabilities to error are as follows: —

1. Impure MnO_2, which sometimes contains C. CO_2 is soluble in H_2O.
2. Solubility of O in water.
3. Escape of gas by leakage.
4. Moisture taken up by the gas.
5. Difference between the temperature of the gas and that of the air in the room.
6. Errors in weighing.
7. Want of accuracy in the weights and scales.

344. Hydrogen.

Experiment 135. — Weigh 5^g or less of sheet or granulated Zn, and put it into a small flask provided with a thistle-tube and a delivery-tube. Cover the Zn with water, and introduce through the thistle-tube measured quantities of HCl, a few cubic centimeters at a time. Collect the H over water in large flasks, observing the same directions as in removing O. Weigh the water, compute the volume of the gas, reduce it to the standard, and obtain the weight, as before. Should any Zn or other solid substance be left, pour off the water or filter it, weigh the dry residue, and deduct its weight from that of the Zn originally taken. Suppose the residue to weigh 0.5^g. Make and solve the proportion from the equation:—

$$Zn + 2HCl = ZnCl_2 + 2H.$$
$$65 \qquad\qquad 2.$$
$$4.5 \qquad\qquad x.$$

Compute the percentage of error, as in the case of O. If the purity of the HCl be known, *i.e.* the weight of HCl gas in one cubic centimeter of the liquid, a proportion can be made between HCl and H, provided no free HCl is left in the flask. State any liabilities to error in this experiment.

PROBLEMS.

(1) A gas occupies 2000^{cc} when the barometer stands 750^{mm}. What volume will it fill at 760^{mm}?

(2) At 750^{mm} my volume of O is $4\frac{1}{2}^l$. What will it be at 730^{mm}?

(3) At 825^{mm}?

(4) At 200^{mm}?

(5) Compute the volume of a gas at 70°, which at 30° is 150^{cc}

(6) At 0° I have 3000^{cc} of O. What volume will it occupy at 100°?

(7) I fill a flask holding 2^l with H. The thermometer indicates 26°, the barometer 762^{mm}. What is the volume of the gas at 0° and 760^{mm}?

If the volumes of gases vary as above, it is evident that their vapor densities must vary inversely. A liter of H at 0° weighs 0.0896. What will a liter of H weigh at 273°? At 273° the one liter has become two liters, one of which weighs 0.0448 ($= 0.0896 \div 2$). The vapor density of a gas is inversely proportional to the temperature. Also, the vapor density is directly proportional to the pressure, since a liter of any gas under a pressure of one atmosphere is reduced to half a liter under two atmospheres.

PROBLEMS.

(1) Find the weight of a liter of O at $0°$; then compute the weight of a liter at $27°$.

(2) Find the weight of 500^{cc} of N_2O at $60°$.

(3) Of 200^{cc} of CO at $-5°$.

(4) A given volume of O weighs 0.25^g at a pressure of 750^{mm}; find the weight of a like volume of O at 758^{mm}.

APPENDIX.

The Author's Laboratory Manual is published by Ginn & Company.

APPARATUS.

Each pupil should be provided with the following apparatus. See frontispiece. Apparatus and chemicals can be obtained of Dr. A. P. Gage, Boston; minerals and metals, of Dr. A. E. Foote, Philadelphia.

4 wide-mouthed bottles (horse-radish size), with corks.
1 soda-bottle.
4 pieces window-glass (3 in. sq.).
2 pieces thick glass tubing (20 in. long, ¼ in. outside diam.).
1 glass stirring-rod.
1 glass funnel (2½ in. wide, 60°).
2 pieces glass tubing (12 in. long, ⅜ in. diam.).
1 porcelain evaporating-dish (3 in. wide).
1 asbestus paper and 1 fine wire gauze (3 in. sq.).

1 iron (or tin) plate.
1 pair forceps.
1 triangular file and 1 round file.
1 copper wire (15 in. long).
6 test-tubes, and corks to fit.
1 wooden test-tube holder.
1 flask with cork (200^{cc}).
1 Bunsen burner (or alcohol lamp).
1 iron ring-stand.
1 piece rubber tubing (18 in. long, ⅜ in. inside diam.).
4 reagent bottles (250^{cc}), HCl, HNO_3, H_2SO_4, NH_4OH.
1 pneumatic trough.

Wherever in this work "Bunsen burner" or "lamp" is mentioned, if gas is not to be had, an alcohol lamp may be substituted.

GENERAL APPARATUS.

The following list includes apparatus needed for occasional use:—

Metric rules (20 or 30^{cm} long).
Metric graduates (25 or 50^{cc}).
Metric graduates (500^{cc}).

Scales with metric weights ($1-200^g$).
Filter papers.
Reagent bottles (250 and 500^{cc}).

206 APPENDIX.

Mouth blowpipes.
Platinum wire and foil.
Mortars and pestles.
Test-tube racks.
Thistle-tubes.
Filter-stands.
Beakers.
Glass tubing ($\frac{3}{16}$ in., $\frac{1}{4}$ in., and 1 in. outside).
Rubber tubing ($\frac{1}{4}$ in., and $\frac{3}{8}$ in. inside).

Hessian crucibles.
Porcelain crucibles.
Electrolytic apparatus, including 2 or more Bunsen cells.
Ignition-tubes.
Steel glass-cutters.
Wire-cutters.
Calcium chloride tubes.
Water baths.
Thermometers.
Barometers, etc.

CHEMICALS.

The following estimate is for twenty pupils: —

Alcohol	1 pt.
Alum	1 oz.
Ammonium chloride	$\frac{1}{2}$ lb.
Ammonium hydrate	1 lb.
Ammonium nitrate	$\frac{1}{2}$ lb.
Antimony (powdered metallic)	$\frac{1}{2}$ oz.
Arsenic (powdered metallic)	$\frac{1}{2}$ oz.
Arsenic trioxide	1 oz.
Barium chloride	1 oz.
Barium nitrate	1 oz.
Beeswax	1 oz.
Bleaching-powder	$\frac{1}{4}$ lb.
Bone-black	$\frac{1}{2}$ lb.
Bromine	$\frac{1}{4}$ lb.
Calcium chloride	1 lb.
Calcium fluoride (powdered)	$\frac{1}{2}$ lb.
Cannel coal	1 lb.
Carbon disulphide	$\frac{1}{4}$ lb.
Chlorhydric acid	6 lb.
Cochineal	1 oz.
Copper (filings)	2 lb.
Copper nitrate	1 oz.
Copper oxide	$\frac{1}{4}$ lb.
Ether (sulphuric)	$\frac{1}{4}$ lb.
Ferrous sulphide	1 lb.
Ferrous sulphate	$\frac{1}{4}$ lb.
Indigo	$\frac{1}{4}$ lb.
Iodine	1 oz.
Iron (filings or turnings)	1 lb.
Lead (sheet)	4 lb.
Lead acetate	1 oz.
Lead nitrate	$\frac{1}{4}$ lb.
Litmus	$\frac{1}{2}$ oz.
Litmus paper	3 sheets.
Magnesium ribbon	3 ft.
Manganese dioxide	2 lb.
Mercurous nitrate	$\frac{1}{2}$ oz.
Nitric acid	3 lb.
Oxalic acid	$\frac{1}{4}$ lb.
Phosphorus	$\frac{1}{4}$ lb.
Potassium (metallic)	$\frac{1}{8}$ oz.
Potassium bromide	$\frac{1}{4}$ lb.
Potassium dichromate	$\frac{1}{4}$ lb.
Potassium chlorate	2 lb.
Potassium hydrate	$\frac{1}{4}$ lb.
Potassium iodide	2 oz.
Potassium nitrate	$\frac{1}{4}$ lb.
Silver nitrate	1 oz.
Sodium	$\frac{1}{2}$ oz.
Sodium carbonate	$\frac{1}{4}$ lb.
Sodium hydrate	1 lb.
Sodium nitrate	$\frac{1}{2}$ lb.

APPENDIX. 207

Sodium silicate	½ lb.	Turkey red cloth		½ yd.
Sodium sulphate	¼ lb.	Turpentine (spirits)		¼ lb.
Sodium sulphide	¼ lb.	Zinc (granulated)		2 lb.
Sodium thiosulphate	¼ lb.	Zinc foil		3 ft.
Sulphur	2 lb.	Sulphuric acid		12 lb.

Additional Material.

These substances are best obtained of local dealers.

Calcium carbonate (marble),	1 lb.	Molasses	1 pt.
Calcium oxide (unslaked lime),	1 lb.	Sodium chloride (fine)	1 lb.
Charcoal	1 lb.	Sodium chloride (coarse)	1 lb.
Sheet lead	4 lb.	Sugar	½ lb.

FOR EXAMINATION.

Those in heavy type are most important.

Rocks and Minerals.

Argillite,
Arsenic,
Arsenopyrite,
Barite,
Calcite,
Cassiterite,
Chalcopyrite,
Chalk,
Cinnabar,
Copper (native),
Corundum,
Dolomite,
Emery,
Feldspar,
Flint,
Galenite,
Granite,
Graphite,
Gypsum,
Hematite,

Hornblende,
Jasper,
Limonite,
Magnesite,
Magnetite,
Malachite,
Meerschaum,
Mica,
Obsidian,
Orpiment,
Pyrite,
Quartz,
Realgar,
Sand,
Serpentine,
Siderite,
Sphalerite,
Talc,
Zincite.

APPENDIX.

Metals and Alloys.

Aluminium,
Aluminium bronze,
Bell metal,
Brass,
Bronze,
Copper,
Galvanized iron,
German silver,
Iron (wrought),

Iron (cast),
Pewter,
Solder,
Steel,
Type metal,
Tin foil,
Tin (bright plate and terne plate),
Zinc (sheet).

Additional Compounds for Examination.

Copper acetate,
Copper arsenite,
Copper nitrate,
Copper sulphate,
Lead dioxide,
Lead protoxide,

Lead carbonate,
Red lead,
Magnesia alba,
Smalt,
Vermilion.

TABLE OF SOLUTIONS.

Number of grams of solids to be dissolved in 500^{cc} of water.

$AgNO_3$	25	$K_2Al_2(SO_4)_4$	50
$BaCl_2$	50	KBr	25
$Ba(NO_3)_2$	30	$K_2Cr_2O_7$	50
$CaCl_2$	60	KI	25
$Ca(OH)_2$	saturated	KOH	60
$CaSO_4$	saturated	Na_2CO_3	50
$CuCl_2$	50	$NaOH$	60
$Cu(NO_3)_2$	50	$Na_2S_2O_3$	saturated
$FeSO_4$	50	NH_4NO_3	50
$HgCl_2$	30	$Pb(C_2H_3O_2)_2$	50
$HgNO_3$	$25 + 25\ HNO_3$	$Pb(NO_3)_2$	50

Other solutions . . . . saturated.

Indigo solution (sulphindigotic acid) is prepared by heating for several hours over a water bath, a mixture of ten parts of H_2SO_4 with one of indigo, and, after letting it stand twenty-four hours, adding twenty parts of water and filtering.

INDEX.

[Numbers refer to Pages.]

A.

Acid reaction, 52.
    salts, 55.
Acids, 45, 49.
    Naming, 50.
Alcohol, ethyl, 184.
Alcohols, 177.
Alizarine, 175.
Alkali, 52.
    belt, 140.
    metals, 52.
Alkaline reaction, 52.
Alloys, 136.
Allotropy, 33, 84, 117.
Alumina, 134, 153.
Aluminium, 151.
    bronze, 137, 151.
Amalgams, 137.
Ammonia, 67, 87.
Ammonium compounds, 145.
    hydrate, 67.
Analysis, 6.
    Water, 44.
Anhydride, 50.
Anhydrite, 149.
Aniline dyes, 179.
Aqua regia, 61.
Argillite, 151.
Arsene, 127.
Arsenic, 126.
    trioxide, 128.
Arsenopyrite, 128.
Atmosphere, Chemistry of, 86.
Artificial stone, 131.
Atom, 6, 9.
Atomic weight, 111.
    volume, 112.
Avogadro's law, 8, 46.
Azurite, 164.

B.

Bases, 45, 51, 67.
Basicity of acids, 55.
Beer, 185.
Bell-metal, 136.
Benzine, 177.
Bicarbonate of sodium, 142.
    of potassium, 143.
Binary, 11.
Bivalent, 39.
Blasting powder, 142.
Bleaching, 99–101, 119.
Bonds, 38.
Bone-black, 34.
Brass, 137.
Bread, 194.
Bricks, 133.
Bromhydric acid, 58.
Bromine, 101.
    compounds, 102.
Bronze, 136.
Butter, 187.

C.

Calcium, 146.
    carbonate, 146.
    hydrate, 70, 147.
    light, 28.
    silicate, 132.
    sulphate, 149.
Carbon, 5, 6, 32.
    allotropic forms, 33.
    a reducing agent, 36.
    a disinfectant, 37.
    an absorber of gases, 37.
    Combustion of, 19.
    dioxide, 19, 79, 87.
    protoxide, 77.

210 INDEX.

Carbonates, 83.
Carbonic anhydride, 80.
  acid, 80.
Carbo-hydrates, 189.
Cassiterite, 163.
Caustic potash, 70.
  soda, 69.
Caves, 148.
Celluloid, 192.
Cellulose, 191.
Centimeters, 1.
Chalcopyrite, 164.
Chalk, 171.
Charcoal, 5, 34.
Chemical activity, 20.
  change, 4, 7.
  union, 9, 13.
Chemism, 5, 6.
Chemistry, 3.
  New theory of, 200.
  of life, 196.
  of vegetation, 123, 197.
China-ware, 133.
Chlorhydric acid, 56.
Chlorine, 98.
Chloroform, 178.
Choke damp, 80.
Cinnabar, 165.
Clay, 151.
Coal, Mineral, 35.
Coal-gas, 182.
Cochineal, 175.
Coefficient, 11.
Coke, 34.
Colloids, 131.
Combustible, 95, 96.
Combustion of metals, 99.
  under water, 122.
Compound, 4, 5, 6.
Condensation, Temperature of, 93, 97.
  of gases, 47, 114.
Conservation, 198.
Cooking soda, 142.
Copper, 164.
  Deposition of, 42.
Copperas, 160.
Corks, To perforate, 15.
Corundum, 151.
Crith, 108.
Cryolite, 105, 133.
Crystalloids, 131.

Crystals, 173.
Cyanhydric acid, 144.
Cyanide of potassium, 144.

D.

Davy lamp, 94.
Deliquescence, 173.
Deoxidizing agent, 36.
Deoxidation in plants, 82.
Dextrin, 192.
Dextrose, 190.
Dialysis, 131.
Diamond, 33.
Dibasic acid, 55.
Diffusion of gases, 114.
Digestion, 194.
Diseases, 195.
Disinfection, 37, 85, 99.
Dissociation, 18.
Distillation, 88.
Distilled liquor, 184.
Divalent, 39.
Divisibility of matter, 3, 4.
Dolomite, 150.
Downward displacement, 98.
Drummond light, 28.
Dyad, 39.
Dynamite, 187.

E.

Earth, A cooling, 199.
  Interior of, 173.
Efflorescence, 173.
Electro-chemical relation, 41.
  table, 43.
Electrolysis, 44.
Elements, 5, 10.
  Table of, 12.
  Per cent of, 173.
Emery, 151.
Energy, 151.
Epsom salt, 151.
Etching glass, 59.
Ethers, 178.
Evolution, 199.
Expert analysis, 127.
Explosions, 27, 96.
Exponents, 11.

F.

Fats, 186.
Feldspar, 151, 172.
Fermentation, 193.
   Lactic, 194.
Fermented liquor, 184.
Ferrous and ferric salts, 159.
   sulphate, 160.
   sulphide, 160.
Filter paper, 3.
Filtrate, 3.
Flame, Chemistry of, 91.
   Bunsen, 92.
   Candle, 91.
   Light and heat of, 93.
   Oxidizing and reducing, 94.
Fluorhydric acid, 58.
Fluorine, 105.
Fluorite, 105.
Food of man, 197.
   of plants, 196.
Fuchsine, 4.

G.

Galena, 161.
Galvanized iron, 153.
Gas-carbon, 34.
Gases, Diffusion of, 114.
   Liquefaction and solidification of, 115.
Gas, Illuminating, 180.
   Natural, 183.
Gaseous molecule, 8, 46, 108.
Gasoline, 177.
Gas volumes and weights, 201.
German silver, 137.
Germs of disease, 87, 195.
Glass, 132.
   Etching, 59.
   manipulation, 14, 15.
Glucose, 190.
Glycerin, 186.
Gneiss, 171.
Gold, 169.
Graduate, 1.
Granite, 172.
Graphite, 34.
Graphic symbols, 39, 176.
Gun cotton, 192.
   metal, 136.
Gunpowder, 144.

H.

Halogens, 106.
   Acids and salts of, 107.
Heat, Absorption and disengagement of, 97.
Hematite, 154.
Hexad, 39.
Honey, 189.
Hydrate, 43.
Hydrogen, 24.
   quantitatively, 202.
   sulphide, 120.
   phosphide, 125.
   sodium carbonate, 142.
Hydrobromic acid, 58.
Hydrocyanic acid, 144.
Hydrofluoric acid, 58.
Hydriodic acid, 58.
Hydrochloric acid, 57.

I.

Ignition tubes, 14.
Illuminating gas, 180.
Infection, 195.
Indigo, 175.
Iodihydric acid, 58.
Iodine, 103.
Iodo-starch paper, 84, 104.
Iron, 154.
   Combustion of, 20.
   rust, 21.
   tetroxide, 20.

K.

Kaolin, 133.
Kerosene, 177.

L.

Lactic fermentation, 194.
Lamp-black, 34.
La Place theory, 199.
Lead, 161.
   compounds, 162.
   Deposition of, 42.
   poisons, 162.
Length, 1.
Levulose, 190.

212 INDEX.

Liebig's condenser, 88.
Lignite, 11.
Lime, 71, 146.
   Superphosphate of, 123.
Limonite, 154.
Litmus, 49, 52.
Lunar caustic, 166.
Luster, 41, 135.

M.

Madder, 175.
Magnesia, 151.
Magnesite, 150.
Magnesium, 150.
Magnetite, 154.
Malachite, 164.
Manipulation, 14.
Marble, 148.
Marsh-gas series, 176.
Marsh's test, 126.
Mass, 4, 5.
Matches, 124.
Matter, Division of, 3.
Meerschaum, 150.
Mercury, 165.
Metals, 41, 135.
Metamorphism, 146.
Metathesis, 7.
Metric system, 1.
Mica, 172.
Microcrith, 108.
Mineral growth, 196.
Minerals, 172.
Mixture, 5, 86.
Molasses, 189.
Molecule, 4, 5, 8.
Molecular differences, 174.
   weight, 109.
   volume, 112.
Monad, 39.
Monobasic, 55.
Monovalent, 39.
Muriatic acid, 57.

N.

Naphtha, 177.
Naphthol-scarlet, 175.
Natural gas, 183.
Neutralization, 53.

Nitric acid, 60.
Nitrogen, 22.
   dioxide, 73.
   monoxide, 72.
   pentoxide, 74.
   tetroxide, 73.
   trioxide, 74.
Nitro-glycerin, 187.
Nitro-hydrochloric acid, 62.
Nitrous oxide, 73.
Non-metals, 135.
Nordhausen sulphuric acid, 66.
Normal salts, 55.

O.

Obsidian, 172.
Oils, 186.
Olefines, 179.
Oleïn, 186.
Oleomargärine, 187.
Organic chemistry, 174.
Orpiment, 128.
Oxidation in the system, 80.
   in water, 81.
Oxidizing agent, 36.
Oxide, 13.
Oxygen, 17.
   a supporter of life, 81.
   quantitatively, 201.
Oxy-hydrogen blow-pipe, 28.
Ozone, 84.

P.

Palmitin, 186.
Parchment, 191.
Paris green, 165.
Peat, 34.
Pentad, 39.
Pewter, 137.
Petroleum, 176.
Philosopher's lamp, 26.
Phosphates, 123.
Phosphene, 125.
Phosphorus, 122.
   Oxides of, 20.
   Red, 124.
Photography, 167.
Pig-iron, 154.
Plant-food, 82, 123, 145, 190, 196.
Plaster of Paris, 149.

INDEX.

Platinum, 168.
Plumbago, 34.
Porcelain and pottery, 133.
Potassium, 143.
    chlorate, 144.
    cyanide, 144.
    hydrate, 70.
Proportion by weight, 29.
    Law of definite, 75.
    Law of multiple, 76.
Prefixes, 13.
Pseudo-triad, 40.
Purple of Cassius, 133.
Putrefaction, 194.
Pyrite, 160.

Q.

Quantitative work, 201, 202.
Quantivalence, 38.
Quartz, 172.

R.

Radical, 40, 43.
Reaction, 25.
    Acid and alkaline, 52.
Realgar, 128.
Reducing agent, 36.
Refractory substances, 134.
Residue, 3.
Rhigoline, 177.
Rocks, 171.

S.

Salts, 45, 53.
    Acid and normal, 55.
    Naming, 54.
Saccharine, 175.
Sandstone, 171.
Saponification, 187.
Saprophytes, 194.
Schist, 171.
Sea-weeds, 104.
Serpentine, 150.
Shale, 171.
Siderite, 154.
Silica, 130.
Silicates, 131.
    Artificial, 132.
Silicic acid, 131.
Silicon, 130.

Silver, 165.
    Salts of, 166.
Slag, 156.
Slate, 171.
Smalt, 132.
Smithsonite, 153.
Soap, 187.
Soda water, 82.
Sodium, 138.
    carbonate, 140.
    chloride, 138.
    hydrate, 69, 142.
    nitrate, 142.
    silicate, 131.
    sulphate, 139.
Soils, 172.
Solubility, 3.
Solution, 3, 4.
Solvent, 3.
Specific gravity, 108.
Speculum-metal, 136.
Sphalerite, 153.
Spontaneous combustion, 96, 97, 125.
Stalactite, 148.
Stalagmite, 148.
Starch, 190.
    test, 103.
Stearin, 186.
Steel, 156.
Sucrose, 189.
Sublimate, 103.
Sublimation, 103.
Sugars, 189.
Sulphur, 116.
    compounds, 119.
Sulphuretted hydrogen, 121.
Sulphuric acid, 63–66.
    Fuming, 66.
Superphosphate of lime, 123.
Supporter of combustion, 95, 99.
Symbols, 10.
Synthesis, 6.

T.

Talc, 150.
Temperature, Critical, 115.
    Lowest, 115.
    of combustion, 93, 97, 152.
    of the body, 81.
Terne plate, 162.

Tetrad, 39.
Theories, 199.
Tin, 163.
Triad, 39.
Tribasic acid, 56.
Turf, 35.
Type-metal, 137.

U.

Union by weight, 29.
  by volume, 46.
Upward displacement, 26.

V.

Valence, 38.
Vapor density, 108.
Vermilion, 165.
Vitriols, 153, 160, 165.
Volume, Metric, 1.
  Molecular, 64.

W.

Water, Chemistry of, 88.
  Electrolysis of, 44.

Water, Hard, 147.
  Pure, 88.
  River, 90.
  Sea, 89.
  Spring, 90.
  in air, 87.
  of crystallization, 139.
Water-gas, 78.
Water-glass, 131.
Weight, 2.
  Least combining, 111.
  of compounds, 29.
White paint, 162.
Wines, 185.
Wood's metal, 137.
Woulff bottles, 56.
Wrought-iron, 157.

Y.

Yeast, 193.

Z.

Zinc and its compounds, 153.
Zincite, 153.
Zylonite, 192.

NATURAL SCIENCE.

Elements of Physics.

A Text-book for High Schools and Academies. By ALFRED P. GAGE, A.M., Instructor in Physics in the English High School, Boston. 12mo. 424 pages. Mailing Price, $1.25; Introduction, $1.12.

THIS treatise is based upon *the doctrine of the conservation of energy*, which is made prominent throughout the work. But the leading feature of the book — one that distinguishes it from all others — is, that it is strictly *experiment-teaching* in its method; *i.e.*, it leads the pupil to "read nature in the language of experiment." So far as practicable, the following plan is adopted: The pupil is expected to accept as *fact* only that which he has seen or learned by personal investigation. He himself performs the larger portion of the experiments with *simple* and *inexpensive* apparatus. such as, in a majority of cases, is in his power to construct with the aid of directions given in the book. The experiments given are rather of the nature of *questions* than of illustrations, and *precede* the statements of principles and laws. Definitions and laws are not given until the pupil has acquired a knowledge of his subject sufficient to enable him to construct them for himself. The aim of the book is to lead the pupil *to observe and to think*.

Wm. Noetling, *State Normal School, Bloomsburg, Pa.:* I know of no other work on the subject that I can so unreservedly recommend to all wide-awake teachers as this.

Benj. F. Thomas, *Prof. of Physics, Ohio State University:* I have used it with preparatory classes for several years with satisfaction. I regard it as the best for class-room work.

H. Wilson Harding, *Prof. of Physics, Lehigh University:* I believe Gage's Elements of Physics to be based on the true method of studying that branch of science, — that of practical work in the laboratory by the student himself.

C. F. Emerson, *Prof. of Physics, Dartmouth College:* It takes up the subject on the right plan, and presents it in a clear yet scientific way.

Introduction to Physical Science.

By A. P. GAGE, Instructor in Physics in the English High School, Boston, Mass., and author of *Elements of Physics*, etc. 12mo. Cloth. viii + 353 pages. With a color chart of spectra, etc. Mailing price, $1.10; for introduction, $1.00.

THE constantly increasing popularity of Gage's *Elements of Physics* has created a demand for an easier book, on the same plan, suited to schools that can give but a limited time to the study. The *Introduction to Physical Science* meets this demand.

In a text-book, the first essentials are correctness and accuracy. It is believed that the *Introduction* will stand the closest expert scrutiny. Especial care has been taken to restrict the use of scientific terms, such as *force*, *energy*, *power*, etc., to their proper significations. Terms like *sound*, *light*, *color*, etc., which have commonly been applied to both the effect and the agent producing the effect, have been rescued from this ambiguity.

Recent advances in physics have been faithfully recorded, and the relative practical importance of the various topics has been taken into account. Among the new features are a full treatment of electric lighting, and descriptions of storage batteries, methods of transmitting electric energy, simple and easy methods of making electrical measurements with inexpensive apparatus, the compound steam-engine, etc. Static electricity, now generally regarded as of comparatively little practical importance, is treated briefly; while dynamic electricity, the most promising physical agent of modern times, is placed in the clearest light of our present knowledge.

The wide use of the *Elements* under the most varied conditions, and, in particular, the author's own experience in teaching it, have shown how to improve where improvement was possible. The style will be found suited to the grades that will use the book. The experiments are of practical significance, and simple in manipulation. The *Introduction* is even more fully illustrated than the *Elements*.

The *Introduction*, like the author's *Elements*, has this distinct and distinctive aim, — to elucidate science, instead of "popularizing" it; to make it liked for its own sake, rather than for its gilding and coating; and, while teaching the facts, to impart the spirit of science, that is to say, the spirit of our civilization and progress.

Alexander Macfarlane, *Prof. of Physics, University of Texas:* I consider that the principal features of the book — its clearness and accuracy of statement, its information being up to date, and the practical nature of the instruction — make it valuable as a first text-book in Physics in high schools and academies, and especially for those institutions that prepare for the universities.

I. Thornton Osmond, *Prof. of Physics, State College, Pa.:* For selection of matter and method of treatment, for comprehensiveness, brevity, clearness, and accuracy, for the simplicity and value of experiments, it was, and yet is, unrivalled as a text-book for high school and academic work.

George E. Gay, *Prin. of High School, Malden, Mass.:* With the matter, both the topics and their presentation, I am better pleased than with any other Physics I have seen.

J. P. Naylor, *Prof. of Physics, De Pauw University:* In its scientific spirit, and in accuracy and clearness of statements of principles, I know nothing that is its superior. The extent to which the work is carried is also about what can be *well* done in the time our schools usually have to give to the subject. It is used in preparatory work at this University as the best we can get.

O. C. Kinyon, *Teacher of Physics in High School, Syracuse, N. Y.:* It not only insures an interest in the study but tends to thoroughly arouse those powers of observation, the development of which is the especial province of scientific study.

B. C. Hinde, *Professor Natural Science, Trinity College, N. C.:* I have used Gage's Introduction to Physical Science for two years, and I consider it the best book published for its purpose. It is strictly in accord with the best modern teaching of Physics. I have made it a point to call the attention of my students to this book that they may use it in their teaching.

Physical Laboratory Manual and Note Book.

By A. P. GAGE, Instructor in Physics in English High School, Boston, and author of *Elements of Physics, Introduction to Physical Science*, etc. 12mo. Boards. xii + 244 pages. By mail, 45 cents; for introduction, 35 cents.

THIS manual has been prepared especially to accompany the author's text-books, but is adapted for use in connection with any good text-book on the subject. The left-hand page contains cuts of apparatus to be used, directions for performing experiments (upwards of one hundred in number), and questions to be answered in connection with the experiments. Suggestions to teachers, the needed tables, etc., are provided at the beginning. The right-hand pages are left blank for the pupil's notes.

A Students' Manual of a Laboratory Course in
Physical Measurements.

By WALLACE CLEMENT SABINE, A.M., Instructor in Harvard University. 8vo. Cloth. ix + 126 pages. Mailing price, $1.35; for introduction, $1.25.

THIS manual, which is intended for use in supplementing college courses in physics, contains an outline of seventy experiments in mechanics, sound, heat, light, magnetism and electricity, arranged with special regard to a systematic and progressive development of the subject. The description of each experiment is accompanied by a brief statement of the physical principles and definitions involved, and a proof of necessary formulae.

Le Roy C. Cooley, *Professor of Physics, Vassar College:* I have examined it and am ready to commend it.

Fernando Sanford, *Professor of Physics, Leland Stanford Junior University:* I like the book very much. It is better adapted to the kind of work which I am trying to do than any other book I have seen.

J. F. Woodhull, *Professor of Science, Teachers' College, New York:* I find Sabine's Laboratory Manual a thoroughly good thing.

High School Laboratory Manual of Physics.

By DUDLEY G. HAYS, CHARLES D. LOWRY, and AUSTIN C. RISHEL, Teachers of Physics in the Chicago High Schools. 8vo. Cloth. iv + 154 pages. Mailing price, 60 cents; for introduction, 50 cents.

THIS manual has been written: First, to present a logically arranged course of experimental work covering the ground of Elementary Physics. Second, to provide sufficient laboratory work to meet college entrance requirements. It contains equivalents of most of the exercises in the Harvard Pamphlet.

The experiments are largely quantitative, but qualitative work is introduced. Apparatus has been chosen that may in most cases be duplicated at small cost. Special care has been taken to make details of work clear, and to instruct the pupil in the methods of making generalizations from his results. Alternate pages are blank for convenience in taking notes.

W. S. Jackman, *Teacher of Science, Cook Co. Normal School, Englewood, Ill.:* It is a most excellent manual and I believe it meets the needs of high schools on this subject better than any other book I have seen.

Introduction to Chemical Science.

By R. P. WILLIAMS, Instructor in Chemistry in the English High School, Boston. 12mo. Cloth. 216 pages. By mail, 90 cents; for introduction, 80 cents.

THIS work is strictly, but easily, inductive. The pupil is stimulated by query and suggestion to observe important phenomena, and to draw correct conclusions. The experiments are illustrative, the apparatus is simple and easily made. Such elements, compounds, and experiments as pupils have no use for, are omitted. The nomenclature, symbols, and writing of equations are made prominent features. In descriptive and theoretical chemistry, the arrangement of subjects is believed to be especially superior in that it presents, not a mere aggregation of facts, but the *science of chemistry*. Brevity and concentration, induction, clearness, accuracy, and a legitimate regard for interest, are leading characteristics. The treatment is full enough for any high school or academy.

Though the method is an advanced one, it has been so simplified that pupils experience no difficulty, but rather an added interest, in following it; the author himself has successfully employed it in classes so large that the simplest and most practical plan has been a necessity.

H. T. Fuller, *Pres. of Polytechnic Institute, Worcester, Mass.:* It is clear, concise, and suggests the most important and most significant experiments for illustration of general principles.

Thos. C. Van Nüys, *Prof. of Chemistry, Indiana University, Bloomington, Ind.:* I consider it an excellent work for students entering upon the study of chemistry.

G. W. Shaw, *Prof. of Chemistry, Pacific University, Forest Grove, Or.:* I am especially pleased with it as filling a place which no other work has filled.

W. J. Martin, *Prof. of Chemistry, Davidson College, N.C.:* I think it is one of the most admirable little text-books I have ever seen.

Wm. F. Langworthy, *Teacher of Chemistry, Colgate Academy, Hamilton, N.Y.:* I am much pleased that we introduced it.

T. H. Norton, *Prof. of Chemistry, Cincinnati University, O.:* Its clearness, accuracy, and compact form render it exceptionally well adapted for use in high and preparatory schools. I shall warmly recommend it for use whenever the effort is made to provide satisfactory training in accordance with the requirements for admission to the scientific courses of the University.

C. F. Adams, *Teacher of Science, High School, Detroit, Mich.*: I have carried two classes through William's Chemistry, and the book has surpassed my highest expectations. It gives greater satisfaction with each succeeding class.

C. K. Wells, *formerly Supt. of Schools, Marietta, O.*: The book bears acquaintance the best of any book of like character that I have ever examined.

W. T. Mather, *Teacher of Science, Williston Seminary, Easthampton, Mass.*: I have used the book in the laboratory very successfully. I can heartily commend it for the method used and the clear and concise treatment of the subject.

J. W. Simmons, *Supt. Schools, Owosso, Mich.*: The proof of the merits of a text-book is found in the crucible of the class-room work.

There are many chemistries, and good ones; but, for our use, this leads them all. There is enough and not too much in the work. It is stated in language plain, interesting and not misleading. A logical order is followed, and the mind of the student is at work because of the many suggestions offered. Our high schools have no province in chemistry beyond the basic facts. Too many text-books go beyond this introductory field, but not far enough to clear away the mists that arise. The student's mind is lumbered with things of which he sees no application. It is not education but the barest kind of stuffing.

We use Williams's work and the results are all we could wish. There is plenty of chemistry in the work for any of our high schools. The above opinion is based upon an experience of twelve years as teacher of chemical science.

Laboratory Manual of General Chemistry.

By R. P. WILLIAMS, Instructor in Chemistry, English High School, Boston, and author of *Introduction to Chemical Science*. 12mo. Boards. xvi + 200 pages. By mail, 30 cents; for introduction, 25 cents.

THE book contains one hundred experiments in general chemistry and qualitative analysis, blanks opposite each for pupils to take notes, laboratory rules, complete tables of symbols, with chemical and common names, reagents, solutions, chemicals, and apparatus, and the plan of a model laboratory. Minute directions, and suggestions designed to help the pupils observe and draw inferences, characterize each experiment.

W. M. Stine, *Prof. of Chemistry, Ohio University, Athens, O.*: It is a work that has my heartiest endorsement. I consider it thoroughly pedagogical in its principles, and its use must certainly give the student the greatest benefit from his chemical drill.

Young's Lessons in Astronomy.

Including uranography. By CHARLES A. YOUNG, Ph.D., LL.D., Professor of Astronomy in the College of New Jersey (Princeton), and author of *A General Astronomy, Elements of Astronomy*, etc. 12mo. Cloth. Illustrated. ix+357 pages, exclusive of four double-page star maps. By mail, $1.30; for introduction, $1.20.

THIS volume has been prepared for schools that desire a brief course free from mathematics. It is based upon the author's *Elements of Astronomy*, but many condensations, simplifications, and changes of arrangement have been made. In fact, everything has been carefully worked over and rewritten to adapt it to the special requirements. Great pains has been taken not to sacrifice accuracy and truth to brevity, and no less to bring everything thoroughly down to date. The latest results of astronomical investigation will be found here. The author has endeavored, too, while discarding mathematics, to give the student a clear understanding and a good grasp of the subject. As a body of information and as a means of discipline, this book will be found, it is believed, of notable value. The most important change in the arrangement of the book has been in bringing the Uranography, or constellation tracing, into the body of the text and placing it near the beginning, a change in harmony with the accepted principle that those whose minds are not mature succeed best in the study of a new subject by beginning with what is concrete and appeals to the senses, rather than with the abstract principles. Brief notes on the legendary mythology of the constellations have been added for the benefit of such pupils as are not likely to become familiar with it in the study of classical literature.

M. W. Harrington, *Chief of U. S. Weather Bureau, Washington, D.C.:* I have been much pleased in looking it over, and will take pleasure in commending it to schools consulting me and requiring an astronomy of this grade. The whole series of Astronomies reflects credit on their distinguished author and shows that he appreciates the needs of the schools.

Clarence E. Kelley, *Prin. of High School, Haverhill, Mass.:* It seems to me the book is admirably adapted to its purpose, and that it accomplishes the difficult task of presenting to the student or reader not conversant with Algebra and Geometry, an excellent selection of what may with profit be given him as an introduction to the science of astronomy.

Blaisdell's Physiologies.

By ALBERT F. BLAISDELL, M.D.

The Child's Book of Health.
Revised Edition. In easy lessons for schools. Illustrated. Mailing price, 35 cents; for introduction, 30 cents.

How to Keep Well.
Revised Edition. A text-book of health for use in the lower grade of schools. Mailing price, 55 cents; for introduction, 45 cents.

Our Bodies and How We Live.
Revised Edition. A text-book of physiology and hygiene adapted for use in advanced grammar schools and high schools. 12mo. Cloth. vi + 403 pages. Mailing price, 75 cents; for introduction, 65 cents.

How to Study Physiology. A Handbook for Teachers. 25 cents.

BLAISDELL'S PHYSIOLOGIES are true, scientific, interesting, and teachable. The matter is fresh and to a considerable extent new. The language is clear, terse, and suggestive. Special emphasis is laid upon the personal care of health. Reference is made throughout the series to the evil effects of stimulants and narcotics on the human system.

The important facts in "How to Keep Well" and "Our Bodies" are illustrated by a systematic series of simple experiments. This feature is peculiar to the Blaisdell books and has been found no less valuable than original. Endorsed by the W. C. T. U.

A Hygienic Physiology.

For the Use of Schools. By D. F. LINCOLN, M.D., Author of *School and Industrial Hygiene*, etc. 12mo. Cloth. Illustrated. v + 206 pages. Price by mail, 90 cents; for introduction, 80 cents.

IT is the distinctive feature of this book to put hygiene first and make anatomy and physiology tributary, instead of making anatomy and physiology the main things and introducing hygiene incidentally.

An Epitome of Anatomy, Physiology, and Hygiene. — Including the Effects of Alcohol and Tobacco.

By H. H. CULVER, formerly Teacher of Physiology in Bishop College, Marshall, Texas. 8vo. Boards. 22 pages. By mail, 25 cents; for introduction, 20 cents. A concise, tabular view of the whole subject.

www.ingramcontent.com/pod-product-compliance
Lightning Source LLC
Chambersburg PA
CBHW031751230426
43669CB00007B/573